Research for Development

Series Editors

Emilio Bartezzaghi, Milan, Italy

Giampio Bracchi, Milan, Italy

Adalberto Del Bo, Politecnico di Milano, Milan, Italy

Ferran Sagarra Trias, Department of Urbanism and Regional Planning, Universitat Politècnica de Catalunya, Barcelona, Barcelona, Spain

Francesco Stellacci, Supramolecular NanoMaterials and Interfaces Laboratory (SuNMiL), Institute of Materials, Ecole Polytechnique Fédérale de Lausanne (EPFL), Lausanne, Vaud, Switzerland

Enrico Zio, Politecnico di Milano, Milan, Italy
    Ecole Centrale Paris, Paris, France

The series Research for Development serves as a vehicle for the presentation and dissemination of complex research and multidisciplinary projects. The published work is dedicated to fostering a high degree of innovation and to the sophisticated demonstration of new techniques or methods.

The aim of the Research for Development series is to promote well-balanced sustainable growth. This might take the form of measurable social and economic outcomes, in addition to environmental benefits, or improved efficiency in the use of resources; it might also involve an original mix of intervention schemes.

Research for Development focuses on the following topics and disciplines:

Urban regeneration and infrastructure, Info-mobility, transport, and logistics, Environment and the land, Cultural heritage and landscape, Energy, Innovation in processes and technologies, Applications of chemistry, materials, and nanotechnologies, Material science and biotechnology solutions, Physics results and related applications and aerospace, Ongoing training and continuing education.

Fondazione Politecnico di Milano collaborates as a special co-partner in this series by suggesting themes and evaluating proposals for new volumes. Research for Development addresses researchers, advanced graduate students, and policy and decision-makers around the world in government, industry, and civil society.

THE SERIES IS INDEXED IN SCOPUS

More information about this series at https://link.springer.com/bookseries/13084

Marco Maria Maiocchi · Zhabiz Shafieyoun

Emotional Design and the Healthcare Environment

All the chapters have been written by both authors, and are related to researches and experiences carried on together, mainly at Politecnico di Milano.

 Springer

Marco Maria Maiocchi
Design Department
Politecnico di Milano
Milan, Milano, Italy

Zhabiz Shafieyoun
Design Department
Winthrop University
Rock Hill, SC, USA

ISSN 2198-7300 ISSN 2198-7319 (electronic)
Research for Development
ISBN 978-3-030-99848-6 ISBN 978-3-030-99846-2 (eBook)
https://doi.org/10.1007/978-3-030-99846-2

This Springer imprint is published by the registered company Springer Nature Switzerland AG
The registered company address is: Gewerbestrasse 11, 6330 Cham, Switzerland

Preface

Causes and Effects in Medicine

Medicine is as old as humans, and many centuries passed before it became a scientific discipline. Looking to the past, but not only, we recognize religious components, sometimes systemic (if not even with cosmological) visions; we can still see such elements today in Chinese medicine, considering the health of a person related to the harmony of the flow of Qi in the whole environment, so that a Chinese doctor must rebalance that flow. This progress went on up to the fifteenth century: in that period a scientific study started, that led to great discoveries.

But we are particularly interested in observing the controversy that arose between Louis Pasteur and Claude Bernard in the mid-nineteenth century.

Pasteur studied external agents and vaccines, and somehow he was assumed as the "champion" of the idea that every disease corresponds to one cause, and that the diagnosis has the goal to identify it, in order to define the correct therapy.

Bernard's view was more related to the presence of a multiplicity of correlated elements, internal and external, which had to be rebalanced, guaranteeing a sort of homeostasis.

Despite the fact that Pasteur obtained more credit at the time, current medicine tends more and more to a systemic vision, very close to that of Bernard, despite the growth of types of specialization.

Recently, during the COVID-19 pandemic period, a friend of us, an Italian doctor (let us say A, male), travelled by car with three other persons (B, C, males and D female); the travel took four hours, and the destination was a working meeting with three further persons (E and F, males and G, female). After the meeting, they returned back, always by car. After few days C was discovered COVID-19 positive and admitted to intensive care.

Subsequent necessary checks verified: A and G negative, B and E positive and admitted to therapy, D and F positive but asymptomatic. The same experience gave

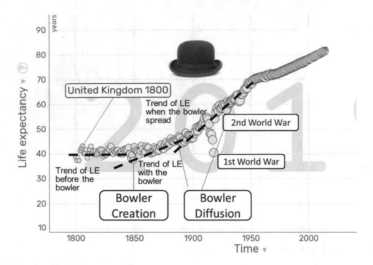

Fig. 1 The use of the bowler since mid-nineteenth century till the end of the twentieth century seems related to the life expectancy (data rearranged from Gapminder—www.gapminder.org). *Source* By Authors

so different effects, because the virus is not a sufficient cause for the contagious nor for the development of the disease.[1]

Moreover, statistical relationships between two phenomena could be complexified by intermediate cause-effect chains.

From Merriam Webster dictionary:

Epiphenomenon:

- a secondary phenomenon accompanying another and caused by it specifically;
- a secondary mental phenomenon that is caused by and accompanies a physical phenomenon but has no causal influence itself.

At the turn of the centuries 19th and 20th, it was recognized a relationship between the use of the bowler (the hut used by financiers in the upper class in London) and the life expectancy in UK; this epiphenomenon was related to the fact that the bowler was used by the upper class of the City, rich and able to access medical health care (Fig. 1).

Again, in 1841 Gustave Flaubert wrote a letter to his sister Caroline, describing the characteristics of a ship (length, height of the mainmast, etc.) and asking to guess the age of the captain. Despite it seems a nonsense problem, and many mathematicians discussed the nature of the question, it is possible to answer with the most likely age of 47 years, due to the fact that the mainmast height can be related to the ship tonnage, and this is in some way a measure of the value of the cargo, and this is related to the required experience for the captain (Fig. 2).

---

[1] The fact, really occurred, has been slightly changed to avoid recognizability of the individuals.

Fig. 2 Relationships between the mainmast height of a ship and the age of the captain (the data have been collected from samples of French harbours registries, but only for explanation purposes, and without any scientific validation). *Source* By Authors

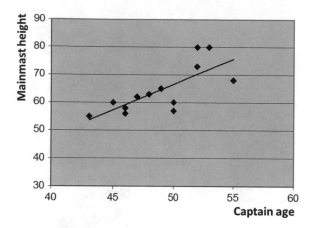

Those two examples show that epiphenomena could lead to get easy but wrong conclusions about possible cause-effect relationships.

So, in medicine, we could have to cope with many epiphenomena, and it could be dangerous to blindly apply the approach of Pasteur.

In fact, application of Pasteur's approach led to many errors, as in the case of scurvy, considered caused by some kind of bacteria, or of toxic substance (the *ptomaine*, whose relationship with scurvy was commonly accepted, but never demonstrated) and so on.

Following Pasteur's approach, doctors verified the existence of a fact (scurvy), accepted without any criticism the idea that this fact had a specific cause and tried to find such a cause to care the "disease". They did not find any cause and "invented" some of them, sure that an external agent was necessary. The diffusion of the "disease" among the crew members confirmed the idea of an epidemic phenomenon related to some bacteria. But someone (e.g. the captain James Cook) observed the reduction of the "disease" in crews eating lemons; they supposed a more complex origin of scurvy (unknowingly following the position of Bernard)[2] (Duesberg, 1996).

The consumption of fresh food, and in particular of lemons, was locally adopted, but the understanding of the relevance of the vitamin C is very late: this vitamin was discovered and isolated only in 1932.

The Relevance of the Boundary Conditions

The presented examples show how complex is the aetiology of a syndrome.

The complex interactions among internal factors (some genetic predisposition, characteristics of the immune system, any intolerances, physical state, etc.) and

---

[2] During an expedition in 1734, a sailor fell seriously ill with scurvy. To avoid the epidemic, he was abandoned on a desert island. Feeding on shellfish and roots, he saved himself, while the rest of the crew died during the voyage (Duesberg, 1996).

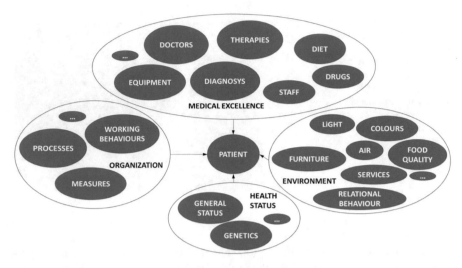

Fig. 3 The complex interaction among the various factors having impact on the health care of a patient. *Source* By Authors

external factors (hygiene, pollution, contamination, healthiness, exposure to contagious or toxic factors, extreme environmental conditions, etc.) make difficult to draw general rules in identifying causes and determine therapies (or, at least, reduce risks).

Moreover, in recent years, research on the nervous system demonstrated the significant dependencies between the processes (and the interactions) within the central nervous system and the peripheral ones (the sympathetic and parasympathetic systems (Sapolsky, 1994), the polyvagal system (Porges, 2011) and so on), and the reactions to pathologies or, in general, to anomalies in health status. *Psichoneuroimmunology* (PNI) is just one of the medical branches studying those aspects (Soresi, 2014).

Nevertheless, the present approach of PNI and similar follow the traditional approach of cause-effect: if there is some psychological problem, it affects the human body system, causing some syndromes and the care consists in removing the psychological cause.

Provided that the health status depends on many boundary conditions, we could not simply increase health by reducing the causes of the disease, but also increasing the positive boundary conditions helping our body, also while influencing our psychological status. So, the context in which we move is highly complex: we can roughly sketch it as follows (Fig. 3).

A patient in a health care environment is immersed in a context influenced by:

– the *medical knowledge excellence*: doctors, the capability of proper diagnoses, the definition of proper therapies, the use of the right drugs, the best equipment and so on, are the main factor of success;
– the general *health status* of the patient, key factor for the healing process: it is a constraint to be properly known;

- the *organization* of the structure: it is the mandatory element for efficiency, effectiveness and economic sustainability, and provides and rules the processes and the behaviours; procedures easy and understandable to patients, reduction of waiting time or empathetic relationships with nurses are examples of elements affected by the organization; all of them can have an impact on patients' emotions and then on the healing process;
- the physical *environment*: it provides stimuli that influence the psychological and the emotional status of a patient and can influence the healing process; also, the behaviour of people can impact on positive or negative relationships influencing the behaviour and the response of the patient.

This book will cope just with the last topics: the *physical environment* and the *organization*. Many aspects will be involved, such as interior design, product design, services, communication and information, wayfinding and so on.

We are aware that the complexity of the global environment creates reciprocal influences and cause-effects chains. It is possible that the relational behaviour of a doctor (e.g. simply by looking in the eyes a patient while visiting) is not confined to the *environment*, because it could influence the personal confidence, improving the compliance in following the prescription, improving the therapy, and then the *medical* aspects and so on.

We cannot provide a general model described in a mechanical way: complexity rules. But we can provide suggestions, elements to take care to, solutions that can be locally adopted, some experiences, and, mainly, a general methodology to carry on a conscious process of improvement, together with a measurement system to provide a reporting useful for a global community.

Milan, Italy   Marco Maria Maiocchi
Rock Hill, USA   Zhabiz Shafieyoun

References

Duesberg, P. H. (1996). *Inventing the AIDS virus*. Gateway Books.
Porges, S. (2011). *The polyvagal theory: Neurophysiological foundations of emotions, attachment, communication, and self-eegulation*. W. W. Norton.
Sapolsky, R. (1994). *Why zebras don't get ulcers*. Henry Holt.
Soresi, E. (2014). *The anarchic brain*. Bookrepublic.

Contents

Chapter 1
Emotional Design

1.1 An Evolutionary View of Design

Following the Industrial Revolution, the industries quickly understood that a strong knowledge of the production process was a key factor: it allowed both to increase the quality of the products and to reduce the production costs. So, it was mandatory to define any element, any characteristic of the product, designing it in any details, and, after that, to set up the proper production process: *Industrial Design* was born.

A clear definition both of all the components of a product and of all the production steps were simple rules to be followed for the purpose; all the producers could apply those rules, putting on the market similar products, with equivalent functions, quality and prices.

So, a competing company had to add to its products some new distinguishing elements. Aesthetic aspects and "meanings" became the differentiating properties: a "nice" product was easier to be sold than a rough one and shapes recalling the "current culture" were more appreciated than anonymous shapes, relevant only for an easier production. Products became not only usable "tools", but a representation of the current culture. Still today, also for a non-expert, it is easy to locate in an historical period a product, just looking to the shapes (see Fig. 1.1). *Design* was born.

But designers can do more: they can astonish the potential customers by fascinating them through emotions, inducing feelings such as *power*, *tenderness* or *desire* (Fig. 1.2). And designers have done it. *Emotional Design* was born.

Meanwhile, the involvement of designers spread in many fields beyond the simple industrial product: communication, interiors, fashion, web, services and so on.

The contribution of Design, and in particular of Emotional Design, is useful also in health care environments. In fact, the impact of this discipline is twofold:

Fig. 1.1 From left: Italian Radio SAFAR of the 1930s, Dumont radio-turntable o the 1950s, radio-turntable by Achille Castiglioni for Brion Vega (1965), Beosound Century by David Lewis for B&O (1993). *Source* Fair use—from Internet

Fig. 1.2 Shapes can provoke arousal of feelings such as aggressiveness (a Lamborghini Miura), softness (a washing machine softener) and lust (advertising by Dolce and Gabbana fashion). *Source* Fair use—from Internet (third photo with reference, as required by license Creative Commons Attribution 3.0, from https://commons.wikimedia.org/wiki/File:%D0%90%D0%B2% D1%82%D0%BE%D0%B1%D1%83%D1%81%D0%BD%D0%B0%D1%8F_%D0%BE%D1% 81%D1%82%D0%B0%D0%BD%D0%BE%D0%B2%D0%BA%D0%B0_-_panoramio_(10). jpg#filelinks)

1.   For the operators (doctors, nurses, staff, etc.), a good design (in the tools, in the organisation, in the furniture, etc.) means a better way of working;

2.   For the patient, a good design (good communication, comfortable furniture, good support and an environment raising positive emotions) means a different approach to illness and a better reaction to the therapies.[1]

We are saying that a good design could improve the effectiveness of care, i.e. quicker healings, less costs, less investments and so on.

---

[1] Many authors support the fact that a good emotional answer of the patient deeply influences the healing process, and the number of papers and books on the subject is growing. We do not list too many references, but simply underline that many approaches supporting the healing process are based the stimulation of positive emotional responses (clown therapy, writing therapy, holistic medicine, music therapy, and so on). Among the others, the popular book E. Soresi (2014). *The Anarchic Brain*, Bookrepublic provides a general framework of the relationships between brain (and emotions) and body (and illness/healing).

Fig. 1.3 A scheme of the
human brain structure.
Source By Authors

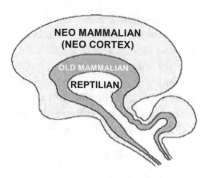

But, in the literature related to Design, the term *Emotional Design* is often used in a generic way, naive and without a well-founded definition.

In the following, we will provide:

a. a scientific definition of the term *emotion;*
b. a model of the human emotional system accepted by the scientific community;
c. the existence of relationships between perceptual "signals" and emotions arousal;
d. the relevance of the metaphoric communication in the process, giving examples.[2]

1.2 Emotional System

Along the species' evolution many physical characteristics were winning for survival: physical strength, fangs, claws, speed in running and so on. Nevertheless, beside the physical aspects, other elements were relevant: behaviours, motivations and triggers.

So, behavioural mechanisms appeared, forcing some type of advantageous reactions to specific stimuli. In front of a weak pray (food) *rage* is useful, but if a bigger and stronger competitor is present, *fear* is safer. So, *rage* suggests attacking, while *fear* suggests running away. Those feelings, *rage* and *fear*, are a new conquest of the evolution, and are what we call *emotions.*

According to most of neuroscientists, the human brain can be described as a composition of three layers, referred to the seniority in the species' evolution. The inner part (*reptilian*) is the site in which primary emotions arise, mainly related to survival (fear, rage, etc.), the middle part, developed in mammals, is related to emotions devoted to the puppies' care and to social relationships, and the upper part (*neocortex*) is mainly related to rational and logic processes (Fig. 1.3).

---

[2] For an in-depth treatment of this subject, see M. Maiocchi, *The Neuroscientific Basis of Successful Design. How Emotions and Perceptions Matter*, Springer, 2015.

Among the various available models describing the physiological mechanisms of emotions, we selected the one developed by J. Panksepp[3] (Panksepp, 2012), accepted by most researchers in the field.[4]

Panksepp, like Darwin (Darwin, 1872), considered emotions as a mechanism "invented" by evolution to increase the survival probability of the species; an emotion is activated in a specific part of the brain, through a specific neurotransmitter class. He defines just seven basic emotions, four of them belonging to the *reptilian* brain, and three to the *old mammalian*.

Belong to reptilian:

- *Seeking*: motivates creatures to explore and become excited when they get what they desire;
- *Rage*: tends to attack as an answer to a possible danger (need of food, competition on the territory or on females, danger for the life, etc.);
- *Fear*: leads creatures to run away, or, when weakly stimulated, to freeze;
- *Lust*: involves sex and sexual desire and allows the species' continuation.

Belong to the old mammalian:

- *Care*: maternal (and, in general, parental) love and caring, protecting babies before they are ready to be autonomous;
- *Panic/Grief*: rules social attachment emotion, from suffering for the absence of maternal care for babies, to the shame, with fear for the rejection by the social group;
- *Play*: pushes young creatures to experience a simulated world, without danger.

All the emotions push for specific reactions through a specific reward: the pleasure (in general coming from the release of dopamine, the neurotransmitter of pleasure).

In addition to the work of Panksepp on the central nervous system, we believe it relevant to consider the broader peripheral system (sympathetic and parasympathetic systems, vagus nerve, etc.), detailed by other authors (Sapolsky, 1994; Porges, 2011). So:

- seven basic emotions arise from the "inner part" of the brain: Seeking, Fear, Rage, Lust, Care, Panic/Grief, Play;
- those emotions can influence the physical state of the entire body (blood pressure, heartbeat, sweat, etc.), that we can connect to emotions/feelings;
- the central and the peripheral systems are in general hierarchically disconnected, and emotions can begin from either of them.

---

[3] According to Panksepp, each emotion rises in a specific part of the brain and is activated by specific neurotransmitters; in this way, it is possible to define scientifically what an emotion is, and to have a well-founded taxonomy of them.

[4] A. Damasio, J. LeDoux and many others (private communication with J. Panksepp).

1.3 Perceptual Stimuli

The complexity of the interactions among the various parts of the nervous system corresponds to the complexity in the arousal of a net of emotions, no matter where perceptual signals come from. The poem *The Raven* by Edgar Allan Poe (1845) is a good example of organisation of many signals coming from different channels (obviously, only described in the poem), concurrently raising emotions related to fear and panic/grief (Fig. 1.4).

The rational part of our brain is far from areas that involve emotion; nevertheless, it has a fundamental role in emotion arousal, being responsible for memory and for analogical and metaphorical reasoning. Consider the two situations depicted in Fig. 1.5; when we encounter a cat having the represented reaction, we feel immediately a risky situation: the cat feels fear and could attack; the knowledge of the risk is not related to a conscious evaluation, but simply induced by primary signals common to many animals (the position of the body, of the ears and the teeth). Without any

"Be that word our sign of parting, bird or fiend!" I shrieked, upstarting—
"Get thee back into the tempest and the Night's Plutonian shore!
Leave no black plume as a token of that lie thy soul hath spoken!
Leave my loneliness unbroken!—quit the bust above my door!
Take thy beak from out my heart, and take thy form from off my door!"
Quoth the raven, "Nevermore."

Fig. 1.4 A strophe from *The Raven* by E. A. Poe. The story (a young man is crying for the death of his beloved woman, when a crow bursts into the room leaning on an ivory bust, answering unconsciously to his desperate questions), the situation (night), the colour contrast (the black raven on the ivory bust) and the deep sound of the repeated words (*nevermore*, recurring at the end of the strophes) create a dark atmosphere and sad emotions, relying on visual, audio and verbal events. *Source* Public Domain

Fig. 1.5 A frightened cat, near to the attack (from C. Darwin, 1872) and Tamara de Lempicka—Autoportrait in a Green Bugatti (1929). *Source* Fair use—from Internet

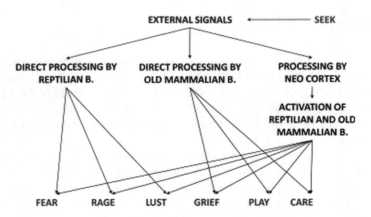

Fig. 1.6 Different emotion arousal paths according to involved brain paths. *Source* By Authors

involvement of the neocortex, we can react changing our path around the cat. The situation is totally different in the case of an elegant woman driving a luxury sports car: in this case, we collect many signals such as the luxury, the dreaming sight, the fashionable dress, the care in the make-up and so on. Our neocortex builds a cognitive structure recalling stereotypical situation related to love affairs and to sex: this logical structure can influence our emotions and then determine our behaviour, and the primary signals are less relevant.

From these examples, it is evident that collections of any type of signals drawn from the environment have relevance, i.e. how relevant *seeking* is; according to Panksepp, seeking is the *mother of emotions*. Seeking is highly relevant because it can influence and raise any other emotion (Fig. 1.6).

1.4   Perceptual Signals

Many communication channels encourage seeking from the external world: viewing, hearing, smelling, touching and tasting, but also pain, heat and so on. A number of studies have been carried out investigating the organisation of our sensorial system, as well as the way in which we react to the various types of signals on specific channels; moreover, recent studies in neurosciences have increased furthering our knowledge.

Those studies also discussed the relationships between the property of a signal and the psychological reaction to it.

Among the various author, we will refer here to the research of Vilayanur Ramachandran (Ramachandran & Hirstein, 1999). These studies are mainly based on visual signals,[5] but in further chapters we will deal with the properties of other perceptual channels, in particular of hearing.

Ramachandran presents the following nine perceptual characteristics, that he verified through Functional Magnetic Resonance Imaging (fMRI), responsible for rising seeking activity.[6]

1. *Peak Shift*: exaggeration of some aspects against the balance of reality; the choice of the enhanced aspects extracts the "truth" from many contingencies; our brain catches quickly the exaggeration and considers it a pleasant meaningful signal (Fig. 1.7).

2. *Perceptual Grouping and Binding*: the human brain tends to group and bind phenomena in order to gather them around a unique already known explanation, corresponding to an abstract shape/figure, in fact not existent (Fig. 1.7).

3. *Contrast*: rods and cones in the human retina are organised to emphasize the perception of edges and contours, i.e. to enhance the *contrast*, distinguishing between background and foreground; the contrast can result from opposition in the colours, but also in a clash in the shapes and in the textures (Fig. 1.7).

4. *Isolation*: this principle refers to the isolation of a single perception modality before amplifying the signal in that modality; for example, we can isolate a shape as first, or also a specific colour (Fig. 1.7).

5. *Perceptual Problem Solving:* our attention is strongly grasped by paradoxical or counter-intuitive situation or structures (think to the Penrose triangle) (Fig. 1.7).

6. *Symmetry:* discovering *symmetry* is rewarding; symmetry is typical in the predators, and our eyes and brain are properly organised to detect it; it is also economy in knowledge, because the property of a part can be used to understand a whole (Fig. 1.7).

7. *Generic viewpoint:* it answers to what designers call *affordance*: being the human visual system driven by Bayesian principles (i.e. driven by the experience), we tend to re-apply already known interpretations, i.e. we like generic point of view and abhor strange coincidences; in this last case, our seeking disagrees, and we feel negative emotions (Fig. 1.7).

8. *Repetition, Rhythm, Orderliness*: our brain, specifically organised to discover regularities, gets comfort in understanding them, end in forecast capability (in space, but also in time). Repetition is also the basis for modularity (Fig. 1.7).

9. *Balance*: the feelings related to this attribute have been examined pointedly and elegantly by Wassily Kandinsky (Kandinsky, 1926) (Fig. 1.7).

---

[5] The principles described by Ramachandran correspond largely to the laws of the *Gestalt* and to the rules largely described by Kanizsa (Kanizsa, 1980).

[6] Again, for an in-depth treatment of this subject, see M. Maiocchi, *The Neuroscientific Basis of Successful Design. How Emotions and Perceptions Matter*, Springer, 2015.

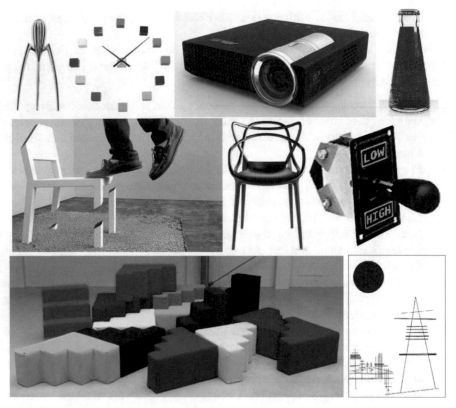

Fig. 1.7 Examples; *Peak Shift*: *Juicy Salif* by Philip Starck for Alessi emphasize the squashing head, deleting as much as possible further details; *Perceptual Grouping and Binding:* wall clock *DIY* by Karlsson; *Contrast*: the projector *P1* by Asus, with contrast in colours (silver colour against black) and in shapes (round against squared); *Isolation*: the original shape of the bottle of Campari Soda (designed in 1932 by Fortunato Depero) is quickly isolated, as well as the distinctive red tonality of the aperitif; *Perceptual Problem Solving*: the *Cut Chair* by Peter Bristol; *Symmetry*: the chairs *Masters* by Philip Starck for Kartell; *Generic View Point*: a violation with a clash between the geometrical position and the meaning in a switch; *Repetition, Rhythm, Orderliness*: the modular multipurpose poufs *Fortyfortwo* by Youngju Oh; *Balance*: a balanced composition according to V. Kandinsky (taking out the disc on the left, the figure results unbalanced). *Source* By Authors

1.5   Metaphors

According to Lakoff (Lakoff & Johnson, 1980), a *metaphor* is an operation based on two different semantic fields, implicitly modelled as entities and relationships. The recognition that the two fields have isomorphic "models" allows us to refer semantically to one field by using verbal references to the other. For instance, if we recognise that the structure of nourishment (a *morsel enters* in the *mouth*, *goes in* the *stomach*, is then *digested*, and provides *vital energy*) is similar to the one of communication (a *word* is *entering* in the *ears*, *goes* to the *brain*, is then *understood*, providing *knowledge*), we can jump between the two semantic fields with phrases such as "I have not digested your words" or "what you are saying is honey to my ears".

We can put in evidence the metaphors by drawing simple semantic networks, such as the ones shown in Fig. 1.8.

A very simple example can show the power of the metaphors: consider the relationships among the sun and common sense (Fig. 1.9, left), in the tragedy *Romeo and Juliet* by William Shakespeare Romeo says (Act III, Sc.2):

> But, soft! what light through yonder window breaks?
> It is the east, and Juliet **is the sun**.
> Arise, fair sun, and kill the envious moon,
> Who is already sick and pale with grief,
> That thou her maid art far more fair than she:
> ...

We are forced to metaphorically associate all the properties of the sun to Juliet: she appears at the window on east, is bright, gives strength, lights up Romeo, feeds him and so on. A single word, *sun*, is able to hurl us in a complex world belonging to our common feelings. Emotions follow.

Not only the words, but also visual structures (and other kinds of signals) can raise metaphoric suggestions: consider the vacuum cleaner *Bidone Aspiratutto* by F. Trabucco (Fig. 1.10): its appearance (the colour army green, the stencils, the

Fig. 1.8 Two different semantic fields can be modelled with isomorphous structures; this allows the construction of metaphorical statements such as *I cannot digest your words* or *I cannot swallow what he is saying*. *Source* By Authors

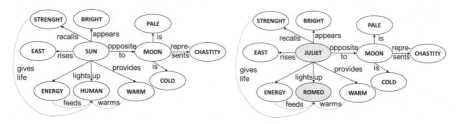

Fig. 1.9 A semantic structure of intuitive, partial, common attributes of the sun and the metaphoric transportation done in Romeo and Juliet by William Shakespeare. *Source* By Authors

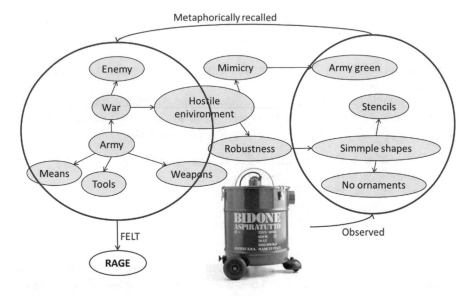

Fig. 1.10 The metaphoric communication of a design product: a vacuum cleaner becomes a powerful weapon for the *marines of the house. Source* By Authors

utilitarian shapes, the big rear wheels and so on) recalls a military environment, and we emotionally feel the *rage* driving the war against the dirty.

1.6 Some Examples in Health Care Environments

Design became a powerful communication mean supporting better and cheaper production processes, able to change the experience of "use" of design artefacts. This notion is enhanced, outside the sensitivity and capability of the designer, by the knowledge of the human emotional system and the mechanisms able to elicit the appropriate emotions and to inhibit the negative ones.

In the following chapters, we will develop a structured model of the areas in which a designer can work in health care environments, to which it is possible to create emotional interventions. We propose a methodology integrated with medical practices, in order to evaluate from a medical point of view the effectiveness of the actions suggested by the designers.

Health care involves many emotional aspects in patients, being they in a negative psychological state of mind related to fear, uncertainty, suffering and so on. Emotional Design is not, of course, a medical solution, but can largely improve the psychological

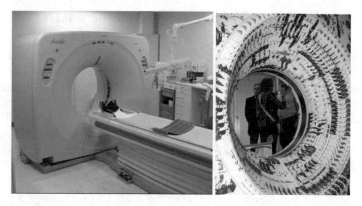

Fig. 1.11 An NMR machine and a similar one "disguised" as a toy. *Source* By Authors

status, and in turn improve the effectiveness of the therapies.[7] We mention here a few introductory experiments, some with measured or qualitatively verified effects to show how wide the intervention area is.

1.6.1 Disguising a Nuclear Magnetic Resonance (NMR) Machine at Children Cancer Centre Pausilipon in Naples

An NMR machine is an imposing object, requiring one to stay motionless for a while, tolerating its annoying noise. The experience is unpleasant for adults and even more for children. To avoid failures in the operations, the children are often sedated, not only for psychological reasons, but also to reduce their mobility. In the Cancer Centre Pausilipon in Naples, a specialized hospital for children, the Radiology Department Chief, Doctor Enzo Salvi, asked the famous artist Mimmo Paladino to disguise the machine, without influencing the operational aspects. Paladino covered all the parts of the machine with funny pictures, creating optical effects, rhythms, intriguing images.[8] The children, during the examinations, activated their *seeking*, reducing *fear* and psychological *grief*. The results have been a dramatic reduction of sedative administration (Fig. 1.11).

Measures indicated that the sedation given to the children fell from 40% to less than 2%: this is a great result, both for the reduction of invasive practices on the children and for the reduction of costs.

---

[7] As previously recalled, many authors (e.g. [Soresi, 2014]) show that the emotional status of a patient can influence dramatically the healing process, as well as the effectiveness of the response to drugs and therapies.

[8] It is easy to recognize the Ramachandran principles of *rhythm* and *repetition*, and the goals of stimulating *seeking* and *play*.

1.6.2   A Reception Desk in a Health Care Environment

Hospitals and Health Care centres typically have a reception desk, in which patients (and/or their families) spend time registering, collecting results or other bureaucratic activities. Usually, a large room with a central desk surrounded by a number of rows of chairs is devoted to these activities. The people waiting are provided with a numbered ticket, which corresponds to information on a monitor indicating their turn. Patients have in front of them a kind of "fortress" to be conquered: they advance towards a convex barrier they need to overcome. No matter what the goal or how well clerks behave: the metaphoric structure is exclusion (Fig. 1.12, left).

The solution used in the Local Health Care Unit *ASL8*, in Asolo (a small town in Northern Italy) is designed differently: there still are tickets for ruling the queues, as well as large screens to inform, and comfortable upholstered chairs too; but what has been changed is the shape of the desk, reverting the curve from convex to concave (Fig. 1.12, centre and right). The emotional result is very interesting: while usually the clerks are perceived as *inside* and the people *outside*, here the contrary is perceived; the desk "embraces" patients. *Inside* and *outside* are powerful primary metaphoric interpretation mechanisms (Lakoff & Johnson, 1980). Signalling protection, refuge and then relieving *grief*: the reverted concavity reduces negative emotions; suggests "hugging" by the desk, advocating *care*.

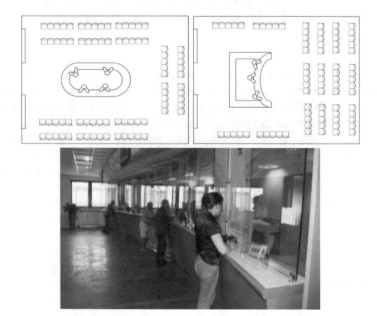

Fig. 1.12 A typical reception desk, optimizing the space for accepting persons and the one used in the ASL8, improving the emotional aspects of care and acceptance. *Source* By Authors

1.6.3 A Breast Radiology Department at Istituto Nazionale dei Tumori (INT) in Milan

The Breast Radiology Department at *Istituto Nazionale dei Tumori* (an eminent cancer centre) in Milan was located in the basement of the building. It is one of the best institutions for the technological equipment, imaging analysis, therapies and for cancer research, but the location was not suitable for the patients' comfort; for this reason, a refurbishment was planned in 2010.

The Department required approximately 400 square metres of floor space, with a reception, waiting room and corridor with many offices and exam rooms.

The protocols for buildings, materials and furnishings for health care centres are well defined strictly enforced. In hospitals, the renovations on buildings are not under the control of the doctors, but of technical offices and regulators. When we were asked by doctors, if they could provide some feedback regarding the refurbishment in order to influence the mood of patients, we were limited and could not accommodate their requests. As a result, we only proposed three changes to be made:

- the colours of walls, doors and furniture;
- the disposition of the chairs in the waiting room;
- many paintings on the walls (Fig. 1.13).

We will describe in detail the experience in the following chapters; however, here we report a sample result along with a recommendation to improve future studies:

- *The result:* the impact on the patients, the doctors as well as the staff working at the hospital has been impressive; the environment was far exceeded the expectation of a hospital. The design of the space triggered the *seeking* activity supplanting patients' health problems, distracting and relaxing them as well as visitors. The

Fig. 1.13 Waiting room at INT in Milano, with orange walls, chairs in purple fabric positioned asymmetrically around the room and many richly coloured paintings on the walls. *Source* By Authors

doctors testified to feel happier and to take a different approach with the patients, and even looked forward to going to work. We also observed a higher frequency of smiling people evident in everyone from that department (Maiocchi, 2010).
- *A recommendation:* from the study, we understood that we had a positive impact but neglected to set up formal procedures to measure it. It was an opportunity to verify through a scientific approach the effects. A longitudinal study would be the most effective in determining the continued effect and account for novelty effects or feelings of newness.

Following this experience, we began to formalize a study to define a process allowing both to measure emotions raised by emotional design actions and to determine the possible influence of emotion on the effectiveness of the therapies.

We carried out several experiments, gaining insights from each study that varied intensity of emotions, including positive and negative effects.

This book is the result of this research and its varied paths.

The following general themes emerged from our analysis:

- the logical and physical structure of a health care environment, including the dynamics of the various processes;
- how emotional design can help in improving the various aspects;
- if it is possible to measure the arousal of emotions and its effect on the healing processes.

Finally, we will provide a general framework for designers to use as a guide for action based on emotional design in the health care environments. Many mandatory codes and requirements are provided by national and regional laws when building new health care centres. We are convinced that those laws should be written in consideration of emotional design principles. The marginal costs of implementing these principles will improve the effectiveness of the healing process, reducing the healing time, which will reduce patient visits, in so doing will reduce overall costs to health care systems.

References

Darwin, C. (1872). *The expression of emotions in animals and man.* Murray.
Kandinsky, W. (1980). *Point, Line to Plane,* Dover. (originally published in 1926)
Kanizsa, G. (1980). *Grammatica del vedere. Saggi su percezione e Gestalt,* Il Mulino.
Lakoff, G., & Johnson, M. (1980). *Metaphors we live by.* University of Chicago Press.
Maiocchi, M. (Ed.). (2010). *Design e Medicina,* Maggioli.
Maiocchi, M. (2015). *The Neuroscientific basis of successful design.* Springer.
Panksepp, J., & Biven, L. (2012). *The Archaeology of mind: Neuroevolutionary origins of human emotions,* W. W. Norton Co.
Poe, E. A. (2002). *Edgar Allan Poe: Complete tales & poems.* Castle Books, (includes the original version of the Raven, published in the *American Review,* 1845)

Porges, S. (2011). *The Polyvagal theory: Neurophysiological foundations of emotions*. Communi-
cation, and Self-Regulation, W. W. Norton & Co.
Ramachandran, V. S., & Hirstein, W. (1999). The science of art: A neurological theory of aesthetic
experience. *Journal of Consciousness Studies, 6*(6–7).
Sapolsky, R. (1994). *Why Zebras don't get Ulcers*. Henry Holt & Co.
Soresi, E. (2014). *The anarchic brain*. Bookrepublic.

Chapter 2
The Goals of Emotional Design for Improving Health Care Environments

It is known that mind, brain and nervous system can be directly and indirectly influenced by elements of the environment. Based on this premise, arises a new field of environmental psychology called "psychoneuroimmunology" (PNI) which, according to Gappell (1991), focuses on the correlation between stress and health. Such discipline promotes a paradigmatic shift in the medicine field: from a rationalist perception of the human body as a group of many different biological systems into a holistic one, where the human body is seen as a strong interrelated connection of bodily and psychological elements. It concerns the interaction between psychological processes, the nervous and immune systems of the human body, through an interdisciplinary approach. Ultimately, it considers how these processes influence health.

Robert Plutchik defines emotion as a complex chain of loosely connected events, the chain beginning with a stimulus and including feelings, psychological changes, impulses to action and specific goal-directed behaviour (Plutchik, 2001); Jaak Panksepp (Panksepp & Biven, 2012) classifies emotions according to specific stimulated areas of the brain and to the neurotransmitter classes involved in their activation. Those definitions lead to a discussion about which kind of action—how and where it is made—will act as a stimulus for a person, inducing a reaction that can affect his or her condition by developing an emotional response to a specific situation.

Reactions provoked by the environment can be positive or negative. There are three primary ways in which the environment can influence patients' outcomes (Rubin, 1998):

- *medical care*: the environment can facilitate or obstruct medical actions of a physician's activities;
- *health status*: the environment can influence the patients' mood which directly affects their condition;
- *causes of the illness*: the environment can expose or protect people from factors that might develop (or worsen) a disease.

M. M. Maiocchi and Z. Shafieyoun, Emotional Design and the Healthcare Environment, Research for Development, https://doi.org/10.1007/978-3-030-99846-2_2

A healing environment must take care of these issues, based on research or evidence, which are used to guide design decisions. Research is based in areas such as environmental psychology, neuroscience and psychoneuroimmunology. A common interest in the research among these fields is the reduction of stress. As the built environment has the potential to be therapeutic, it must be designed to reduce environmental stress factors that might negatively affect the users. It is important to note that, in the field of health care facilities, stress affects not only patients but also personnel.

Stress can be generated by several factors; illnesses and treatments might cause *Fear*, bring uncertainty, impair physical capabilities, expose patients to painful and frightening procedures, expose physicians to high-tension operations and so on. All these facts are stress stimulants, but stress can also be generated by noisy and unwelcoming spaces. It can affect immune function through emotional and/or behavioural manifestations (such as anxiety, fear, tension, anger and sadness), and physiological changes (such as blood pressure, heart rate and sweating).

There are two pathways by which our bodies respond to stress. The former regards the endocrine system. This system secretes hormones such as *cortisol* that, once inside the bloodstream, is distributed to different organs, entering the cells and provoking changes in their function. Elevated levels of cortisol impair the immune system (Rabin, 1999). The latter regards the autonomic nervous system, which can stimulate the sympathetic nervous system and thus the secretion of *adrenaline*, a hormone which causes blood pressure to rise and can also affect different bodily systems, as it happens with cortisol.

Stress is a natural human reaction and an important one. Historically, stress has been a useful tool in keeping humans alert and safe from harm. In the short term, it can actually be positive. In the long term, however, stress response weakens the immune system's ability to resist disease. The body's response turns negative, starting at the cellular level and then broadens to immune functions, resulting in the suppression of immunity (Sapolsky, 1994). Therefore, it is of great importance to reduce stress in order to fortify and stimulate the immune system, contributing for the achievement of health.

Considering that the environment can impact people's mental and emotional states and that, as psychoneuroimmunology stands, our nervous system is intrinsically connected to our immune system. It could be said that a "positive" environment can affect the mind and emotions of individuals in a good way; it will also positively affect their nervous system and therefore the immune system. Further, an environment unavoidably communicates with people and produces emotional effects. Such an environment creates information that will be relayed to the brain. That information will, turn, initiate a series of connections within the body. These connections are able to produce substances that affect the immune system and ultimately the patient's health (Fig. 2.1).

Having understood that stress hinders the immune system, and that the environment can present factors that emotionally affect people in a positive way (leading to the reduction of stress as one of the outcomes), designers must think of the environment as psychologically supportive. The purpose of space shifts to promoting patient

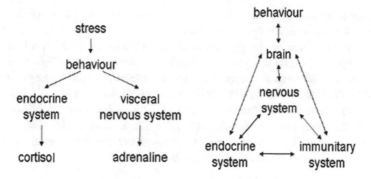

Fig. 2.1 Pathways for stress and behaviour and impacts on endocrine and immunity systems. *Source* By Authors

empowerment and enabling patients, relatives and staff to cope with and overcome illness.

A health care facility is a complex system, and therefore, it is important to have well outlined organizational and operational protocols along with the built space. "Even the most well-designed physical setting cannot be considered a healing environment without hospital operational protocols that empower patients and staff to accommodate individual preferences and beliefs" (Malkin, 1992).

Improvements we can consider refer to:

- elements measurable as physical quantities;
- reaction to therapies (time):

  - healing process (time);
  - compliance to prescriptions (communication);
  - ...

    all of them will be measured directly:

- elements that do not correspond to physical quantities:

  - trust in doctors;
  - trust in the organization;
  - trust in the therapies:
  - ...

    most of them will be measured through Kansei Engineering (see next chapter).

2.1 History

For a long time, we have understood the power of emotion on behaviour. Darwin's view was one the first to emphasize the role of emotion in shaping social behaviours and human communication in his theory of *evolution* (Darwin, 1872). He focused on expressions in his book *Expression of Emotion in Man and Animal*. He demonstrates that different facial and body expressions are under the influence of emotions such as surprise, anger, hate, anxiety, grief, fear, joy, love as well as others. He argues human emotions and behaviours are universal, and they are not related to the environment or culture differences. One hundred forty-nine years later we are still talking about emotions and their vital role in understanding people. Emotion might not only be the response to an external stimulus, but it can influence experience (Dewey, 1934). Then, to understand the emotion of people, we need to consider external and internal stimuli, human experience and environmental effect.

2.2   Perception—Emotion—Behaviour—Environment

From the moment we are born the first things we see starts to form our perception. People's perception creates emotions by interoperating sensations as the first step in overall thoughts. Thoughts are hidden, but the emotional responses express themselves in different ways. People's facial and body expression of different emotions might be the same (Ekman, 2003), but their reaction to the emotion and their behaviour are different. Someone might yell at someone when they are angry, while the other person might stay silent or leave the scene. External factors including noise, colour, temperature, services as well as other people's behaviour can be perceived differently causing a variety of emotional and behavioural responses.

According to Bruce Goldstein, perceptual process is a sequence of stimuli, electricity, experience, action and knowledge working together to create our experience and reactions to stimuli in the environment (Goldstein & Brockmole, 2016). Reassessing Goldstein's process, we can account for perception as a response to stimuli as everything we sense around us vying for our attention. Electric impulses are triggered making connections between our senses and our brain. [Re]actions follow from the perceived state; recognition and action caused by stimuli. We bring our experience to the perceptual situation as well mitigating our reaction through tried electrical impulses. Environmental stimuli can be anything around us, anything we see, smell, hear, touch or taste. Sensation, perception, emotion, behaviour and the environment continuously affect each other creating our varied experience. Then, to create a good experience we need to pay attention to the process of perception, especially the beginning which establishes our first impression. The first impression of a hospital initiates an experience, which has an indelible effect on future impressions and trips to the hospital. It is critical to remember that changes to stimuli impact emotion, behaviour and experience as well. Designing to stimulate or to please, introducing constraints, rules, comfort (but also pleasure) triggers emotions.

Environmental factors are able to persuade specific behaviours, such as motivating people to lead a healthier life, by influencing their perception and emotion (Smith et al., 2017). The role of environmental design on people's experience and their outcome is undeniable. A hospital needs to create an environment to improve people's emotion ultimately improving their treatment outcomes.

2.3 Information as a Stimulus

Hospitals can be a complex mix of information, locomotion and interaction. A "patient journey" can be confusing with lack of information about different steps of the journey. Information is one of the stimuli to trigger stress or create *Care*.

Information related to each patient journey includes the length of time spending in each phase, procedure of their treatment, the equipment's and the way are facing with it, the rules and effects of medicine and so on.

Knowing the information perceives *Care* and remove the stress of being in an unknown place with an unclear procedure. Knowing is power, and our brain wants to control the future experience. Losing ability to control things make people unhappy, hopeless, helpless and depressed (Gilbert, 2009). By getting sick people are losing a part of the power of control, something happens to them out of their hand, even people want to control uncontrollable things also. They need to have better control on the rest of the process, treatment, recovery and going back to the normal life. Lack of information about their journey adds to their feeling of losing control and give them the feeling being controlled. The role of information in patient care is very important, people should feel they are a part of the group of decision-makers about their illness, their body and their time. Information should transfer clear, understandable and with respect to create an easier life in the hospital.

2.4 Patient Journey

2.4.1 The Arrival

Arriving to the hospital is one of the most stressful part of the journey for patients, especially for new patients. Needs for *Care* is starting by the first step of being in the hospital. Patients want to know where they are, what they should do, where they should go and how all the process would be and how long does it take.

Usually, patients or their companies are seeking for some signs to show them the path some people to welcome them and answer their questions.

According to Norman (2013), we can consider three levels of design: *visceral*, *behavioural* and *reflective* (borrowed by the ABC model in psychology. Affective, Behavioural and Cognitive components):

- *visceral design*: the affective component; perceptible qualities (shapes, colours, styles and so on) can recall feelings able to unconsciously and automatically influence emotions (e.g. hard lines vs smooth curves, related to scary vs relaxing mood);

- *behavioural design*: related to usability; in general, how easy and pleasant the use of an object or the interaction with an environment are;
- *reflective design*: refers to cognitive aspects, and to beliefs and thoughts we associate with an object or an environment. So, we can consider something technologically innovative or related to luxury, representing or not the one's self-image and so on.

Visceral design is more to emotional aspects. Considering the patients' journey, visceral design in arrival can be related to some signs or map of the hospital, to get their attention, and a reception or information desk near the arrival.

First impression of cleanness of the hospital, smell and the way of welcoming is very important to create a good first experience of the hospital.

Smell of the hospital, for example, is an environmental stimulation that might interpret the place dirty or clean. Some people have a memory of special smell of the hospital which interpret it to sickness, pain and grief.

Creating the new experience needs new stimulations and emotions. First impression of smell, noises, human behaviours, colours and furniture should perceive supportive, clean, trustable and respectful to create positive emotion and behaviours. The emotion of people changes during the consuming design, and even after the consumption, and it could have effect on their loyalty (Chitturi, 2009). People emotions change during their journey, at the end of the journey or even few years after. If a hospital wants to keep the customers loyalty, to refer the hospital to others and they come back again if they need it, watching these changes and sending them home satisfied will be important. Following people in their journey and after their journey to respond to their needs in a right way is a solution. Responding to the need of each design level, visceral, behavioural and reflective is the key to keep the level of the people's emotions positive.

After arrival patients should go to the check-in point to register and wait to visit the doctor. The things they see, touch, hear or smell on their way to the waiting room have direct effect on their perception of the hospital and the quality of care. They might cause feeling stress and anxiety, or care and joy, or combination of them. Stress drives from variety of situation, physical, social, biological and chemical that can be triggered by internal or external sources (Gunnar & Barr, 1998).

Knowing what is going on in people's mind it is not easy, but we can operate on external sources (i.e. the physical environment situations such as colour, furniture layout, temperature, light, noise, other people's behaviour). According to (Ulrich et al., 2004), nature and indoor plants are decreasing the stress of people, using plants or the nature atmosphere can be a solution to make the journey more pleasurable.

2.4.2 Check-in and Waiting Rooms

After arriving to the waiting room, it is very important to be able to talk to an operator without any hesitation in asking questions and making people confident of being in a safe place. Most of the time receptionist are too busy, or they are overloaded with work, and they might not have enough time to spend with the patients who might interpret it as a disrespectful behaviour, causing stress or panic for patients. Decreasing the nurses' workload to take care of a smaller number of patients in each shift and training them before start working with patients will decrease the amount of stress for both staff and patients.

Check-in for new patients takes longer: it will give them the stress of losing their time and they might not fill the forms carefully. Operators should make them sure about their time and let them to fill the forms in their seats, when they have their numbers in their hands or they are sure they don't lose their turn because of answering the questions in filling the forms. The operator behaviour has the most emotional impacts on patients: smiley faces with a careful use of words to interoperate, time, care and support make them comfortable. Answering fast and get them to be quick will be perceived as obligation and as losing control. Negative emotions such as grief or panic start to arouse and immune system start to be affected. Check-in for new patient and old patient both should go smoothly: one inappropriate question might have a bigger effect in people who are suffering from a sickness. Knowing about the emotion of patients in each stage of a treatment helps operators to understand them better. For example, people at the first stage of their treatment might be more confused and stressed than at the end of their treatment.

Waiting rooms are one of the most stressful part of the hospital (Altringer, 2010). Patients' perception of time increases in long wait (Camacho et al., 2006) (Anderson, 2010). Patients always feel other patients are getting more attention before them, continuously they check the time and numbers. They choose their seat in front of the reception to be seen. They are trying to be aware to everything because they believe there are not enough care for them. IDEO, as a design and consulting firm, in of the health care project for the *De Paul Medical Center*, made waiting room relaxing with more comfortable furniture, adding vinyl wood floor and repainting the walls to a soft hue (Altringer, 2010). A comfortable waiting room might strongly reduce the stress of the long wait, but still there are some anxiety and pain to be solved. According to our studies at Istituto Nazionale dei Tumori in Milan (Italy), sometimes patients do not care about the attraction of physical objects and environment, and they need more attention and information. Their stress would decrease if they see a smiley face or someone to share their worries with them. Combining the relaxing environment with a good service, such as clear information about time to visit the doctor, will help. Calling people closer to them by their name, breaking the long wait to the smaller waits in different places. Furniture layout provides sitting alone or in a group. The main waiting room in INT had a glass roof and it was letting people to feel themselves in a bigger place; seeing the sky was helping them to see the nature and feel less stress. Most of the people interviewed in this waiting room had no complaints about the environment, but they had questions about their situation, and they wanted someone to give them hope.

2.4.3 The Visit

2.4.3.1 The Examination Room

From the patient's perspective, the reward for a long hospital or clinic wait is to see someone who knows all the answers to your questions; the doctor. This part of the journey is one of the most important, because it can mitigate all the stress of the previous steps. Spending a large amount of the time with the doctor and asking questions increases the quality of care dramatically. Doctors know that their job is to build trust, how to give patients hope, or make them scared if they are not listened to, or not following their prescribed advice. The role of the doctor along the journey is pivotal but is situated within the context of their location. A doctor's behaviour within the atmosphere of the doctor's office too can project an air of authority, care or any number of traits to satisfy a patient's journey. Di Blasi and others aim to reduce negative feelings such as anxiety and fear in their study. They go on to suggest, physicians with a warm, friendly, and reassuring behaviour with their patients are more effective in increasing positive feeling (Di Blasi et al., 2001). In addition to doctors and staff behaviour and environment have emotionally effect on patients. Zaisel, Silverstein and Levkoff contend behavioural health outcomes in Alzheimer patients that specially designed facilities can help reduce negative emotions (Zeisel et al., 2003). They go on to show that proper lighting can affect circadian rhythms and the result on brain functions are positive. Benefits of nature or access to natural environments are positive for patients can also be seen in medical studies (Ulrich et al., 2004). Design solutions such as providing windows that reveal natural vistas and their effect on improving medical outcomes are demonstrated in Roger Ulrich studies showing the effect of the natural environment on hospital patient outcomes is undeniable (Ulrich et al., 2004). Anne Di Nardo (2013) in Health Care Design magazine discusses using aromatherapy with natural scents of lemongrass in a Brazilian Hospital, Mãe de Deus resulting in a general overall improvement in patient outcomes. Additionally, it increases the family and staff satisfaction during their time in the hospital as well.

In addition to the doctor's demeanour and the office they work in, the examination itself is an interaction that triggers and builds on the patient's experience. Examinations bring patients and doctors closer both physically and in their relationship of trust. Patients put their trust in the doctor to do everything necessary to resolve the issue. During the patients most vulnerable point in the journey, the role of the doctor becomes more sensitive not only to the patient's medical needs but also to their emotional state as well. Doctors should explain each step of their examination, before they begin and during the examination to assure the patient of their actions. Patients should know the reason for the examination; the equipment doctor uses and the effect on the patient. Expectations of the patients can range from what the equipment does, to what is the temperature of the device? Is it going to be painful? And what kind of pain should I expect and what is abnormal?

Sometimes before a doctor's visit, the patient will be asked to remove their clothes and laydown on the examination table; they need to know how long they will be exposed and who else may be entering the room? Who can see into the room when the door is opened and how long the examination will take while they are in this vulnerable position? In some test rooms patients may be in this position for 20 to 30 min without any idea when or how long they will need to wait. This wait is perceived to be longer than the wait before seeing the doctor in the waiting area, because it is the purpose of their visit, and their expectations are higher. The levels of anxiety are even higher in anticipation of moving beyond the diagnosis towards treatment. Often patients understand that the problems they are experiencing are symptoms of underlying malfunctions. What follows the doctor's examinations is inevitably negative news regarding to their conditions. The combination of dread and anticipation of hearing about their status effect the perception of time and their ability to live through it as normally they would.

2.4.3.2   From Diagnosis to Tests

Performing tests, like every other part of the journey in the hospital, is a process that is unknown to the general public. Patients have to deal with the emotional aspect of the diagnosis and at the same time find their way through hallways and corridors on towards the testing room. At the same time, they need to understand instructions on what to prepare for and how the procedure will unfold. For example, different tests such as X-ray, CT scan, MRI and blood tests have different procedures and needs some preparation before they begin.

Patients should prepare mentally and physically for each step of the process. During x-rays for example, it depends on which part of the body needs to be scanned exposing their limbs or removing their clothes. They must remember to remove metal items such as jewellery. If a chest x-ray is needed patients must put on a hospital gown, expose their chest and navigate the rear closure of the open back. The issues of the hospital gown are well documented from the variety of materials to the difficult rear tying of the three-point string closure. Patients already feeling vulnerable due to their condition feel equally vulnerable from their body being exposed during the test or just feeling uncomfortable wearing a gown with an open back. Patients who are new to this process may need more explanation. Most of the time staff give them the instruction very quick and leave them alone in a change room to figure out the process by themselves.

From the preparation to the procedure, a patient will be left alone in an x-ray room during the exam. Typically, a technician will talk to them through a speaker to indicate when to hold your breath and when to resume breathing. We can imagine the sterile feeling of an x-ray room with white walls, being spoken to from a speaker concerned for one's health, dressed in the most minimal of clothes. The experience is more similar to entering prison or beginning basic training in the army than someone who should be cared for while they are sick. Patients should be shown the process before having the x-ray, verbal explanations or visualizing the process using aids such as flyers or a video playing in the change room can be used to ease fear and anxiety.

CT scans have a similar preparation and process of changing cloths, removing any metal and exposing the body parts needing to be scanned. Sometimes CT scans require an injection during the test which can react differently with each patient to the medicine, and they should be watched closely for further changes. Most of the time the medicine makes people feel warm, and patients should expect these changes in the body, so they do not get alarmed when they occur. CT scan machines can be scary if people do not have enough information beforehand. The scanner looks like a narrow tunnel with a mobile bed within it. The bed moves slowly inside the tunnel during the exam to scan different areas of the body. Some people with claustrophobia might have problems during the test or may experience this sensation for the first time. There are little technicians can do to move or react to their fear during the test other than to stop scanning. Any movement by the patient will have an effect on the test and will also require repeating it. Often patients are provided with a call button to push if they do not feel well during the exam. If they learn about the machine prior to the test or experience the non-threating space with professionals watching may assure them they are safe, reducing levels of stress.

MRI machines are similar to the CT scanners with a few small differences. The shape of CT scanner is described as "donut" shaped, and an MRI machine is more like a "tanning bed". The length of MRI is longer than CT scanner. An MRI might take between 30 and 45 min, but CT is approximately 5–15 min. This means patients stay longer in the MRI tunnel which can lead to stress. MRI machines make loud and unusual noises not likely heard by patients before. The noises could be perceived as annoying, tedious and persistent making patients anxious. Some hospitals offer a type of headphones to reduce the noise and focus on something else, but they also help people hear the operator better. Without headphones people might not be able to tolerate all the noise and again request to leave the test incomplete. Patients should be supported before the test helping to experience the noise differently for themselves.

During a blood test, patients may see that the nurses are taking more blood than needed for the test, which might make some people nauseous or make them suspicious about their health and procedures of the test. An image of a chair and a sharp needle in the vein with a number of full test tubes can seem more like a horror movie than a place for people's health. Learning about people's emotions while seeing sharp equipment is very important: we naturally perceive this as a threat. Bar and Neta (2007) demonstrate how sharp objects trigger a threat code in the brain. The designers of such equipment should reconsider their designs, avoiding "sharp products" and "disguising" them towards a visceral design: doctors and nurses will use it in such a way that patients will not be scared and will not interpret it as threatening. Attempts have been made to keep needles out of sight from patients as much as possible. Providing some information like how much blood must be taken for their test, and why this is necessary, as well as the expected recovery needed can decrease the amount of stress during the test.

Patients under stress due to uncertain situation are forced to deal with a number of unknowns, what is perceived as advances in technology by professionals, the casual observer only sees a black-box of a device soon to invade their body. Providing knowledge to people makes machines less threatening and changes the experience of working with machine to mitigate the stressful experience. On the other hand, each explanation takes additional time prolonging the visit as well it becomes repetitive information for the technicians and nurses to give. For these reasons, information must be delivered both efficiently and to save time for the patients. This can be done again using video or interactive tools that covers any questions about the test, decreasing the number of retelling instructions for nurses. In more extreme cases using virtual reality (VR) may be another solution to put patients at ease, simulating the experience to prepare patients for the test. The audiences who suffer from anxiety and when the process is complex enough this may be a viable option provided these simulations can be shared and used in a variety of settings.

Giving information to patients before doing their test is often interpreted as providing care. Patients need to know they are safe and if anything appears threaten to them, the hospital is ready to help. Nurses and doctors are trained to encourage feelings of safety for patients by showing their attention and concern before, during and after the test. Before the test, patients need to know the type of treatment and what they should expect; moreover, they need to know that caregivers are with them, or watching them even when they are alone in a room. During the test, patients need to feel safe and secure. After the test, nurses should be careful about their reaction and their words they use to make sure they do not inadvertently worry the patient. Patients always think nurses or doctors will see something in their test immediately following a test monitor every reaction, gesture and facial expression. Nurses must assume the results need some investigation, and nothing is clear or can be explained at this point before a consultation with doctors and specialists.

2.4.4 Patient's Results

Waiting for the results, particularly for patient with life threatening diagnosis, are the most stressful stage of the journey. Hospitals not only strive to make this stage as short as possible but also need to balance the information speed with accuracy to provide the best prognosis. If patients are able to track their results, they stay engaged in the process helping them to have less stress awaiting the test outcome. At the stage of diagnosis, there is a tendency to still be optimistic even if the evidence is dire. Thoughts of minor causes compounded problems that can be easily explained or new treatments that provide quick answers flood the mind of many patients. Following tests, however, the factual, scientific and reality of the situation are often solidified. In serious cases, the anguish and fear of the diagnosis become apparent and little else can occupy the mind. Only time can bring the patient to begin to process their test results. At this stage in the process, the perception of time can slow or skip by without notice. Unfortunately, time may not be afforded if decisions need to be made and procedures need to be planned to deal with the condition. Patients always have many questions after getting their result, and it is better to answer these questions immediately after they receive them to both empower the patient with information but also continue to engage patients in their own treatment. Calling patients or giving them the result in person is also perceived as providing care and not just providing information. For example, if they need to see another doctor or keep a special diet or have a follow-up exam scheduled, these should happen quickly to decrease worry and anxiety. During the result stage of a patients' journey, waiting for information is the most critical and time should seem to move a determined pace in order to present confidence in a plan that moves a patient from the result stage to the stage of treatment and on to recovery.

2.4.5 Follow-up Visits

Discharge from the hospital after getting the result is not the end of the patient journey. Patients need to book additional appointments for follow-up visits. Follow-up procedures are visits, treatment, therapies or other consultations about the patient's condition' Follow-up journey should happen easier and shorter than first hospital journey. The follow-up patients can be shorter by reducing some part of it like check- in, patient information is already recorded at the first journey, and there is no need to fill a lot of forms about their insurance or demographic information. Doctors already have their patient's medical records and patients are familiar with the hospital structure. Hospital should be supportive with follow-up appointment with remind patients to book their follow-up appointment, remember their appointment and explain the benefit of follow-up to them. They need to feel secure about their next journey in the hospital. Patients feel Care by reminding them about their next visit or their treatment.

Centres for Medicare and Medicaid Services in 2010 proposed providing patients with clear discharge instructions and appointments for timely follow-up visits to reduce avoidable hospital readmissions (Grafft et al., 2010).

2.4.6 Re-admittance

According to the New England Journal of Medicine, readmission rates for targeted conditions (ailments for which patients are seeking treatment) declined from 21.5% to 17.8% citing the changes brought on by the affordable care act of 2007 (Zuckerman et al., 2016). Although this is good news for re-admittance, a nearly 18% re-admittance rate could still be reduced to cut costs and improve overall patient care. Re-admittance is caused by a number of expected factors including failure to understand discharge orders, poor care, early discharge, lack of communication after discharge, unclear verbal or written instruction, not finishing the entire dosage of prescribed medicine, no follow-up appointment, mall nutrition, socio-economic status, obesity, ethnicity and gender, language Zapatero (Zapatero et al., 2013), (Canadian Institute for Health Information, 2012), (Metzger, 2012). Re-admittance is costly and become an important outcome for evaluating the quality of care (Mah et al., 2019). It is also clear from the NEJM study that there are many reasons that can cause re-admittance even from unforeseen circumstances related to the previously mentioned reasons. For example, mall nutrition makes the body week which has a related effect on understanding the treatment requirements or interpreting the doctor's use of language differently or being confused once a patient is at home. Designing better communication tools to discharge patients will assist with their treatment outside the hospital. A study in IIT Institute of Design the health care design team created a discharge tool "to improve asthma self-management skills and outcomes" (Erwin et al., 2016). This tool made it easy for patients to learn how to selfcare after discharge from the emergency room. A flyer showing them the instruction visually and clearly encourages patients to recognize the actions they need to take rather than recall what the doctor had told them. Looking at images showing the follow-up treatment is more helpful at the moment is open to selfcare than remembering the doctor's instructions or reading some notes. These instructions also need to be suited for the patient mode of receiving information, such as paper booklet or flyer, internet website, or phone app. Sending patients home with the right directions in the context they prefer is another way to decrease re-admittance causes.

2.5 Emotion

Emotional considerations are the key to healing patients and at the same time improve the health care system. Showing empathy during all forms of therapy makes it more effective. Studies have shown that empathic physicians with warm and friendly manner have a placebo effect on patient's therapy, speeding recovery and effectiveness (Larson & Yao, 2005). In a study in 1993, Dale and Anthony use different forms of interview questions to understand what they termed, emotional "connexion" (A rapport between patients and clinician). They identified four features that clinician responses to patients' emotions were required and made explicit to convey empathy; acknowledgement, support and partnership, legitimation and touch. These four are meant to elicit the emotion of being cared for from the patients. These are minor in terms of time requirements but massively beneficial towards improvement. Instead of enquiring in a direct way such as, "did the pain travel to your neck?" using acknowledgements like, "that must have been painful if it has moved into your neck". To build support and partnerships with patients a clinician may respond, "It's important to me to understand your fears, so we can address them together". Together with acknowledgement, legitimating a patient's concern is also a feature found in empathy. Statements such as "Who wouldn't be afraid after something like that?" can recognize that medical professionals have the same human responses. Although more socially charged in current pollical climates done compassionately, a simple physical gesture can transfer emotion instead of using words (Matthews et al., 1993). Eric Larson believes making "connexions" and building empathy are principal to caring and enhance the therapeutic potential of patient clinician relationships (Larson & Yao, 2005).

2.5.1 Emotional Design

Emotional Design definition by Lo is the focus on the need and experience of the user. Emotional design through the function, form and usability enhances the user experience (Lo, 2007). Emotional design includes design a service, product or experience, supports emotional experience of patient care. There are many examples of effect of environment on patient's experience in the hospital and the hospital outcome. IDEO is one the companies with many projects in health care, including Kaiser Permanente (KP) and DePaul. IDEO design responded to the emotion of people by creating some design solutions such as putting whiteboard in patient's room to let the visitors (family and friends) leave a caring message (Altringer, 2010). The benefit of these projects influences the patient care experience and the treatment outcome (Altringer, 2010).

2.5.2 Emotional Architecture

Despite still underway, we want recall here the project *NuArch*, in progress at the dell' Istituto di Neuroscienze del Consiglio Nazionale delle Ricerche (IN-C.N.R.) in Parma (Italy). The project aims to measure how the shape of space can modify the emotional content of people's experiences and to investigate the more complex aspects of the relationship between the shape of space and the bodily and affective cerebral representations of humans.

The research is carried out within virtual reality environments, able to activate the neurophysiological responses of the entire architectural perception in a social scenario.

Some preliminary results (Presti, Ruzzon, Avanzini et al., 2021, Presti, Ruzzon, Caruana et al.,2021) show that:

- shrinking walls provoke a higher level of activation in the subjects than architectures in which the walls remain constant or widen;
- windows located in a higher position cause in the subjects a higher level of activation than architectures in which the height of the windows remains constant or decreases;

- shrinking walls are less pleasant than those in which the walls remain constant or widen;
- widening side is preferred to stay longer than those with constant walls or with decreasing width;
- the colour of the walls is less significant in affecting the emotional perception of the architecture.

References

Altringer, B. (2010). The emotional experience of patient care: A case for innovation in health care design. *Journal of Health Services Research & Policy, 15*(3), 174–177.

Anderson, D. (2010). *Humanizing the hospital: Design lessons from a Finnish sanatorium.* CMAJ.

Bar, M., & Neta, M. (2007). Visual elements of subjective preference modulate amygdala activation. *Neuropsychologia, 45*(10), 2191–2200.

Camacho, F., Anderson, R., Safrit, A., Jones, A. S., & Hoffmann, P. (2006). The relationship between patient's perceived waiting time and office-based practice satisfaction. *North Carolina Medical Journal, 67*(6), 409–413.

Canadian Institute for Health Information. (2012). All-cause readmission to acute care and return to the emergency department. *Health System Performance.*

Chitturi, R. (2009). Emotions by design: A consumer perspective. *International Journal of Design, 3*(2), 7–17.

Darwin, C. (1872). *The expression of emotions in animals and man.* Murray.

Dewey, J. (1934). *Art as experience.* Penguin Putnam.

Di Blasi, Z., Harkness, E., Ernst, E., Georgiou, A., & Kleijnen, J. (2001). Influence of context effects on health outcomes: A systematic review. *The Lancet, 357*(9258), 757–762.

Di Nardo, A. (2013). Rethinking behavioral health center design. *Healthcare Design Magazine.*

Ekman, P. (2003). Darwin, deception, and facial expression. *Annals of the New York Academy of Sciences, 1000*(1), 205–221.

Erwin, K., Martin, M. A., Flippin, T., Norell, S., Shadlyn, A., Yang, J., Falco, P., Rivera, J., Ignoffo, S., Kumar, R., Margellos-Anast, H., McDermott, M., McMahon, K., Mosnaim, G., Nyenhuis, S. M., Press, V. G., Ramsay, J. E., Soyemi, K., Thompson, T. M., & Krishnan, J. A. (2016). Engaging stakeholders to design a comparative effectiveness trial in children with uncontrolled asthma. *Journal of Comparative Effectiveness Research, 5*(1), 17–30.

Gappell, M. (1991). *Psychoneuro-immunology.* 4th Symposium on Healthcare Design, Boston, MA.

Gilbert, D. (2009). *Stumbling on happiness.* Vintage.

Goldstein, E. B., & Brockmole, J. R. (2016). *Sensation and perception.* Boston Cengage learning.

Grafft, C. A., McDonald, F. S., Ruud, K. L., Liesinger, J. T., Johnson, M. G., & Naessens, J. M. (2010). Effect of hospital follow-up appointment on clinical event outcomes and mortality. *Archives of Internal Medicine, 170*(11), 955–960.

Gunnar, M. R., & Barr, R. G. (1998). Stress, early brain development, and behavior. *Infants and Young Children, 11*(1), 1–14.

Larson, E. B., & Yao, X. (2005). Clinical Empathy as Emotional Labor in the Patient-Physician Relationship. *JAMA, 293*(9), 1100–1106.

Lo, K. P. Y. (2007). Emotional design for hotel stay experiences: Research on guest emotions and design opportunities. In *Proceedings of International Association of Societies of Design Research*

2007 Conference: Emerging Trends in Design Research, The Hong Kong Polytechnic University, Hong Kong.

Mah, J. M., Dewit, Y., Groome, P., Djerboua, M., Booth, C. M., & Flemming, J. A. (2019). Early hospital readmission and survival in patients with cirrhosis: A population-based study. *Canadian Liver Journal, 2*(3), 109–120.

Malkin, J. (1992). *Hospital interior architecture: Creating healing environments for special patient populations.* Wiley.

Matthews, D. A., Suchman, A. L., & Branch, W. T., Jr. (1993). Making connexions: Enhancing the therapeutic potential of patient clinician relationships. *Annals of Internal Medicine, 118*(12), 973–977.

Metzger, J. (2012). *Preventing hospital readmissions: The first test case for continuity of care.* Global Institute for Emerging Healthcare Practices.

Norman, D. (2013). *The design of everyday things* (Revised and expanded ed.). Basic books.

Panksepp, J., & Biven, L. (2012). *The archaeology of mind: Neuroevolutionary origins of human emotions.* W. W. Norton Co.

Plutchik, R. (2001). Integration, differentiation, and derivatives of emotion. *Evolution and Cognition, 7*(2), 114–115.

Presti, P., Ruzzon, D., Avanzini, P., Caruana, F., Rizzolatti, G., & Vecchiato, G. (2021). *The experience of virtual environments affects the perception of emotional body postures: an adaptation after effect pilot study.* Annual Conference of the Society for Affective Science.

Presti, P., Ruzzon, D., Caruana, F., & Vecchiato, G. (2021). *Architectural forms' impact on perceived valence and arousal of virtual environments.* Symposium of the Academy of Neurosciences for Architecture.

Rabin, B. S. (1999). *Stress, immune function and health: The connection.* Wiley Liss.

Rubin, H. R. (1998). Status report-an investigation to determine whether the built environment affects patients' medical outcomes. *Journal of Healthcare Design, 10*, 11–13.

Sapolsky, R. (1994). *Why Zebras don't get ulcers.* Henry Holt & Co.

Smith, M., Hosking, J., Woodward, A., Witten, K., MacMillan, A., Field, A., Baas, P., & Mackie, H. (2017). Systematic literature review of built environment effects on physical activity and active transport–an update and new findings on health equity. *International Journal of Behavioral Nutrition and Physical Activity, 14*(1), 1–27.

Ulrich, R. S., Zimring, C., Joseph, A., Quan, X., & Choudhary, R. (2004). *The role of the physical environment in the hospital of the 21st century: A once-in-a-lifetime opportunity.* The Center for Health Design.

Zapatero, A., Barba, R., Ruiz, J., Losa, J. E., Plaza, S., Canora, J., & Marco, J. (2013). Malnutrition and obesity: Influence in mortality and readmissions in chronic obstructive pulmonary disease patients. *Journal of Human Nutrition and Dietetics, 26*, 16–22.

Zeisel, J., Silverstein, N. M., Hyde, J., Levkoff, S., Powell Lawton, M., & Holmes, W. (2003, October). Environmental correlates to behavioral health outcomes in Alzheimer's special care units. *The Gerontologist, 43*(5), 697–711.

Zuckerman, R. B., Sheingold, S. H., Orav, E. J., Ruhter, J., & Epstein, A. M. (2016). Readmissions, observation, and the hospital readmissions reduction program. *New England Journal of Medicine, 374*(16), 1543–1551.

Chapter 3
A Model of a Health Care Organization and Environment, System Design and Interior Design

3.1 Towards a Model of a Health Care Environment

This book copes with three different disciplines: Health Care, Design and Neurosciences. In particular, we want to approach Design interventions in Health Care environments, according to neuroscientific principles.

For this purpose, we need to set up a model of a Health Care environment, in order to be able to classify the possible actions.

We will not discuss at all about medical issues.

A hospital, a medical analysis laboratory and a First Aid centre have different roles and structures, but we can model all of them through a formal abstract representation, allowing us to provide a general structure suitable for the description of each of them.

A Health Care centre can be described as shown in Fig. 3.1.[1]

It is depicted as a "box" having, as input, *sick people* ("entering" in the "box" to be properly treated), and, as output, *healed people* and an increased *medical knowledge*.

The box is provided with an "engine" (the bottom arrow), composed of the *medical* staff (doctors and nurses) and of *administrative staff*, working in a *physical environment* (a building, with rooms, laboratories and offices, and with all the equipment and furniture), according to properly defined *procedures*, and thanks to the *medical knowledge*.

The function of the structure is constrained by various elements, among which *legal and ethical rules* (but also by economic limits, medical knowledge limits and so on).

The structure is dynamic and evolves: for example, the treatments provide data useful for the growth of the *medical knowledge*, so that the output results can affect the system.

---

[1] We use the formalism IDEF0, derived from the established graphic modelling language Structured Analysis and Design Technique (SADT) developed by Douglas T. Ross and SofTech, Inc. (see *Systems Engineering Fundamentals*. Defense Acquisition University Press, 2001).

© The Author(s), under exclusive license to Springer Nature Switzerland AG 2022
M. M. Maiocchi and Z. Shafieyoun, Emotional Design and the Healthcare Environment,
Research for Development, https://doi.org/10.1007/978-3-030-99846-2_3

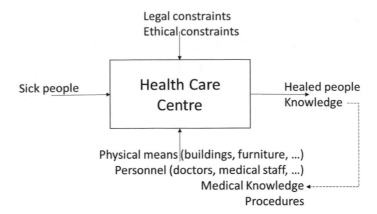

Fig. 3.1 Abstract model of a generic Health Care centre. *Source* By Authors

The model is roughly abstract (e.g. we neglect the relevant factor "money", the drugs, the relationships with authorities and so on), but it is sufficient for our purpose.

3.2   Action Areas

Besides the obvious different roles of patients and doctors and nurses, there is another relevant peculiarity differentiating them: while doctors and staff are totally aware of the rules and of the location, patients require to be driven (physically in the space and in behaviour) and should be constantly informed on the rules. We underline this point because a scarce information on what to do induces uncertainty, i.e. Seeking becomes unsatisfactory, producing stress.

This fact reveals that, besides the perceptual aspects of the physical elements, we must consider also the functional ones (such as procedures, expected behaviours and so on).

Again, procedures are mandatory for a proper management of all the processes, but the way in which those are implemented can affect patients' emotions. For example, the check-in procedure in a hospital can be ruled either with a queue management as in a post office or as at the reception of a luxury hotel; the emotional impact in the two cases will be dramatically different: the customer of a post office feels to be just "a number", a task to be completed; the customer in the luxury hotel feels to be a welcome guest.

Finally, interaction between patients and personnel is crucial and goes far beyond procedures: for example, a doctor looking into the patients' eyes during a visit communicates *Care*.

According to the above considerations, the actions driven by emotional design should refer to:

- Physical elements, such as:

  - Building architecture;
  - Interior materials and colours;
  - Lights;
  - Furniture;
  - Equipment;

- Functional elements, such as:

  - Signs;
  - Information panels;
  - Explicative monitors;

- Procedures
- Behaviours.

3.3 A Journey Through a Medical Experience

All elements above must be matched with the different situations occurring in a Health Care centre. We consider the following steps as a typical journey through a medical experience:

- *Registration*: the first action that connects patient and health care centre, establishing a relationship between them; it can be an appointment, a request for a visit or control, or more;
- *Doctor visits*: typically providing a first diagnosis, or the needs of further exams/investigations;
- *Exams*: usually related to samples and laboratory analysis, or by equipment (ultrasound, X-ray, PET, MNR, etc.)
- *Diagnosis*: the response defining the therapy to be applied;
- *Check-in*: registration for hospitalization;
- *Treatment/surgery*: the therapy;
- *Life in health care centre*: refers to any activities possible apart from the medical aspects (reading, TV watching, people visits, entertainment, etc.)
- *Doctor visit*: periodical checks;
- *Check-out*: discharge from the health care centre (sometimes for definitive healing, sometimes because hospitalization is no more required).

For each of the above steps, we can consider specific characteristics: for example, signs are more important for reaching exams laboratories, waiting rooms will be present for exams, as well as for the check-in, and so on.

The journey is just is only a presentation expedient, used in order not to neglect any of the possible situations. Obviously, the various passes and treaties may also be valid for single actions and single visits that do not involve hospitalization.

We recall that, for each of them, our aim is to rise positive emotions (Seeking, Care, Play) and to reduce negative emotions (Rage, Fear, Grief). We can reach our goals by taking into account the perceptual principles by V. Ramachandran exposed in Chap. 1 and the possible metaphorical interpretation of the environment.

3.4   A Map of Emotional Design Actions

We can build a Table 3.1 for a complete map, as reported below.

The map is organized according to a possible complete "journey" in a hospital, starting with the registration and ending with the check-out.

As already indicated, we use this map of a journey just as a way to orderly discuss all the elements, but a patient will get advantages also for a single indicated step: for example, a visit by a doctor could involve only the registration, the visit, without any further step, but the considerations we will do are valid also for those two steps.

For each of the possible steps, we will consider the four intervention areas, i.e. the physical layout, the functional aspects, the procedure ruling the functioning of the hospital and the behavioural code of the doctors and staff.

For each of those areas, we indicate either the elements on which we can apply emotional design principles, or, when it is difficult to be indicated with precision, the goal we intend to reach. For example, it is easy to list the possible elements in which we can apply principles of emotional design for the physical environment (e.g. furniture, interiors, signals, etc.), but it is not so evident which elements could be affected for the functional aspect or for a procedure. For example, the functional aspect during a visit cannot simply be related to the proper diagnosis, but must involve the emotional status of the patient. Furniture and environment can have an impact on the functions from ergonomics point of view, but we need more: to enhance the patient's awareness, without provoking Fear or Grief, but rising Care; this can be obtained in different ways (speeches by doctors, booklets, videoclips, but also the behaviour of nurses and staff, etc.). In such cases, we will list the goals we would like to reach.

The same apply to procedures and behaviours.

In some parts of the table, some goal such as *efficiency* appears: of course, efficiency is a general goal for each organization, coping with economic aspects; we intend here a different interpretation; when we speak about efficiency we intend that the speed, the precision and other related parameters can affect the emotional status of a patient, and this is our goal; efficiency can involve the reduction of waiting time, the correspondence between the expectations of the patients and the facts, and so on, which can increase the trust in the organization, positively affecting the emotions.

Let us examine the meaning and the goals of each element appearing in the cells of the table.

Remember that our overall goal is to:

Table 3.1 Map of Emotional Design actions along the "journey" of a patient

	Physical elements	Functional	Procedures	Behaviours
1. Registration	Layout Interior Furniture Signs Waiting rooms	Layout Furniture Signs Helpdesk	Awareness Status Efficiency	Reputation Relationship
2. Visit	Layout Interior Furniture Signs Waiting rooms	Awareness	Efficiency	Reputation Relationship
3. Exams	Layout Interior Furniture Equipment Signs Waiting rooms	Awareness	Awareness Status Efficiency	Reputation Relationship
4. Diagnosis	Layout Interior Furniture	Awareness		Reputation Relationship
5. Check-in	Layout Interior Furniture Signs Waiting rooms	Awareness	Awareness Status Efficiency	Reputation Relationship
6. Treatment	Layout Interior Furniture Signs	Awareness	Awareness Status	Reputation Relationship
7. Life	Layout Interior Furniture Signs	What to do Outdoor		
8. Visit	Layout Interior Furniture Signs	Awareness	Awareness Status	Reputation Relationship
9. Check-out	Interior Furniture Signs	Awareness		Relationship

- increase Seeking, Care, Play;
- reduce Fear, Rage, Panic-Grief.

1.   *Registration*

The first contact with a Health Care centre imprints an emotional evaluation difficult to be changed in the following.

The patients are entering the centre with fear for their illness and should find an environment suggesting a warm welcome and *Care*; as a person unaware of the place, the *Seeking* activity should be easily satisfied, allowing them to locate what is needed. So,

- the layout should be easy to be identified and modelled, both physically (shape of the area, without hidden corners, and a clear separation of the spaces) and functionally (functions available in the various spaces) (*Seeking*);
- warm colours (*Care*) could guide in the interpretation of the areas (*Seeking*);
- lighting should be used to emphasize the organization in different areas (*Seeking*);
- paintings/posters and pictures in general on the walls can provide the impression of a lively environment, possibly affecting *Play* and reducing *Panic-Grief*;
- coloured furniture with smoothed shapes can increase *Care*;
- clear signs, easy to be perceived and interpreted, will increase *Seeking*, as well as the presence of a helpdesk;
- because it is possible that the patient has to wait in line, it is relevant a ticket distributor, as well as monitors informing not only the call, but also the status (position in the queue, expected time to wait, etc.), to guarantee the awareness of the process and the status, so increasing *Care* and reducing *Panic-Grief*; of course, besides the awareness on the status, the processing speed is relevant: procedures and organization should be efficient;
- the reputation of the centre boosts the confidence in it, but reputation is obtained also with the described actions; the patients trust in structures offering *Care*;
- finally, it is very important the relationship between personnel and patients, based on empathy: it is possible to suggest to personnel to be kind and to smile in the interactions, but the background for such a behaviour is a positive working atmosphere and a solid employee satisfaction.

The registration can start also through a telephone call or an on-line procedure. In this case, the same aspects we examined above arise, but in a different context.

The physical environment is substituted by screenshots implementing the procedures, and the Emotional Design affecting interiors, furniture and signage must be replaced by a good Web Emotional Design.

In such cases, due to the reduction of the human–human interaction, efficiency and awareness are mandatory.

Efficiency means that the customer should get reliable information in a very short time and get the services in a properly predicted (and short) time.

These topics too are presented and discussed in paragraph 5.2.

2. *Visit*

The first visit is a relevant event: the patients will know the origin of their illness, the therapies to be followed, the prognosis. This moment is dominated by uncertainty and Fear.

After the patients had an appointment (typically in the Registration step), they must go in the Health Care centre and reach the doctor's office, following the proper pathway. A proper signage (as shown in Sect. 5.2) will increase *Seeking* and reduce *Fear*.

At the office, the patient will be welcomed by a nurse and made them to sit in a waiting room. Waiting is a recurring situation in the patients' journey: for the first visit, for any exams; moreover, the permanent mood along the hospitalization is to wait: patients live a suspended situation in respect to the normal life. For this reason, the topic related to the waiting rooms will be addressed separately, with a specific chapter dedicated to them.

Finally, the patient enters the doctor's office, for the visit.

The typical office appears white, with white lacquered metal cabinets and chairs, creating an aseptic environment, cold and suggesting absence of emotional signals: a place in which illness is breathed.

It is possible to change the appearance of furniture and interiors, always respecting the health standards, adding colours, paintings on the walls, and, in general, to equip the office as a home. So, avoiding to recall continuously the illness, we reduce *Fear* and inspire *Care*.

Of course, the behaviour of the doctor has a bigger impact: to look the patient into the eyes, to smile, to explain everything during the visit largely increase *Care*.

3.   *Exams*

There are many elements emotionally influencing a patient during exams. Beside the waiting rooms (see Sect. 5.3 and following), signage to reach the proper exam room (see Sect. 5.2), interior and furniture (see Sect. 5.3), there are other peculiar facts.

A laboratory is a specialist environment, with complex and sophisticated equipment, operated by experts, and often to be operated carefully, due to possible danger (e.g. X-ray machines). Moreover, some exams last just few minutes while others (such as NMR or Radiotherapy) can take more and more time. Finally, often the doctors work with some monitor, preferring half-light. Also, a proper initial response, transparent and understandable to non-experts, as patients are supposed to be, is relevant to reduce *Fear* and *Panic-Grief*. So:

- the layout of the laboratory should avoid any possible interpretation of separation between doctor and patient (Fig. 3.2);
- equipment has often incumbent structures (f. i. NMR or Radiotherapy machines), and some kind of disguise, if possible, could be helpful (Fig. 3.3);
- useful distractions for the patient during the exams or therapies could be helpful too (Figs. 3.4 and 3.5);
- the behaviour of the doctor is very important: looking into the patient's eyes and cautious smiling communicate *Care*;
- the patient must have the information about the results of the exam as soon as possible, and, in any case, some indications should be given immediately after the exam. Moreover, the awareness of the mechanisms of the disease, as well as of the role of exams and of the mechanisms of the therapies, increases *Seeking* and *Care* and reduces *Panic-Grief* (Figs. 3.6 and 3.7).

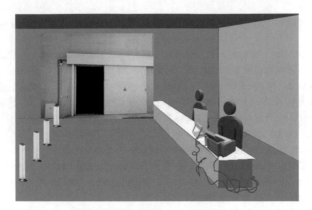

Fig. 3.2 A typical situation for accessing a Radiotherapy session (sketched from a real case). The patient must follow a corridor as a path towards a "torture"; nurses are observing, but are separated by a fence, are on the other side, looking as spectators, without participation (this kind of communication is based on a metaphoric interpretation of the layout); a different layout, locating equipment on the wall and making nurses share the area with the patient could mean *Care* and reduce *Fear* and *Panic-Grief*. *Source* By Authors

Fig. 3.3 Above, a detail of the disguised NMR equipment at the Pausilipon Hospital (already described in Chap. 1). Below, three different proposals done by students of the Instituto Tecnológico y de Estudios Superiores de Monterrey (ITESM) (Queretaro, Mexico) for some children's hospital in Mexico. The works were co-ordinate by the professor Cynthia Ortega del Castillo (Maiocchi, 2007). *Source* By Authors

Fig. 3.4 *Imaginary Journey*, a work by the artist Gian Antonio Garlaschi. Those spheres have been hung on the ceiling of the Radiotherapy room at Istituto Nazionale dei Tumori di Milano. During the therapies, the patients can observe them slowly moving, and, according to the movements of the medical equipment, from different points of views. *Source* By Authors

Fig. 3.5 Doctor Enzo Salvi, director of the Radiology Department of the Children's Oncologic Centre Pausilipon of Naples shows a portable ultrasound equipment, joined with a monitor connected to a DVD player; young patients can attend to a movie while doing the ultrasound, increasing *Play* and *Care*, and reducing *Fear*. *Source* By Authors

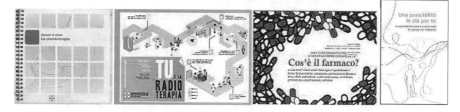

Fig. 3.6 Some actions to increase patients' awareness at Istituto dei Tumori di Milano. Informative booklets oriented to the description of the various diseases and therapies. In the figure, the booklet about Radiotherapy, with the visualization of the parcourse to be followed for the single application, the cover of a booklet about drugs, the cover of a booklet on the possibility to participate to experimental researches. These booklets play a relevant role also for the following steps, in particular for *Treatments*. *Source* By Authors

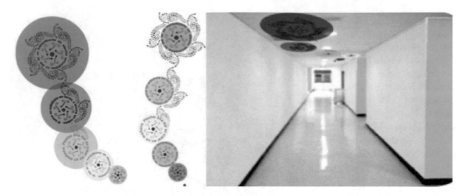

Fig. 3.7 Also the passage from one department to another is a special moment. When patients are moved from their rooms to a visit or exam, they are often placed on a gurney or couch and are "guided" by nurses to the destination. At that moment, the patients see only the uniform ceiling of the corridors, without even a spatial reference of where they are going. A proposal made at *Clinica 33* in Queretaro (Mexico) provides paintings of multicoloured "mandalas" on the ceilings, shapes rich in symmetries and intriguing doodles, which, if suitably differentiated, can also constitute a valid spatial reference. The patient not only gets distracted, but also avoids the sense of loss that could lead to the onset of *Panic-Grief*. *Source* By Authors

4.   *Diagnosis*

A testimony

> Following a small traumatic event that caused me an annoying and persistent pain in the shoulder joint, an orthopaedic specialist requested me to take an X-ray; I choose to go at a specialized diagnostic centre. Following an efficient telephone booking, I went there, I easily paid my "health ticket", then I am given a set of printed sheets and a satisfaction measurement questionnaire (to be returned at the end of the visit, filled in); then I was sent to the Radiology area. Here, a sub-receptionist welcomed me, smiling, and giving me a folder, equipped with a small display, and inviting to sit in a waiting room, and to wait for a signal from the same folder. Within about fifteen minutes a beep from the folder attracted my attention: I read to on the display the number of the room to go. I approached, knocked, and entered a small essential study, in dim light. The doctor, bended over his computer, took the documentation I offered him, invited me to put myself shirtless and to sit on a bed next to an ultrasound equipment; then, with his gaze always focused on the screen, he carried out the ultrasound, memorizing the images he considered most relevant. Then he invited me to get dressed, dictated to the computer the description of the results and the diagnosis, verified the correct transcription, and absently told me about an alteration of an articular surface. "Doctor, I congratulate you on the level of technology you are equipped with!". A 'thank you' closed the dialogue and the visit. I never saw that doctor's eye colour. I went out with my report form in my hands and, before going out, I tore up and threw away the satisfaction survey form.

Communicating a diagnosis is an extremely relevant moment for the patient, particularly if the diagnosis indicates the existence of a problem. Environmental context design can do very little if the doctor is unable to empathically communicate with the patient, infusing him with a sense of his authority, trust in care and hope for recovery.

Fig. 3.8 Check-in at Niguarda Hospital (Milano) (photo by Sbragagia from Wikipedia) and Check-in at Vier Jahreszeiten Hotel in Munich (Germany). *Source* First: https://it.wikipedia.org/wiki/File:Ospedale_Niguarda-area_sala_attesa_per_accettazione.jpg with license Creative Commons Attribuzione-Condividi allo stesso modo 3.0 Unported. Second:https://commons.wikimedia.org/wiki/File:Front_desk-hotel.JPG - I, the copyright holder of this work, release this work into the **public domain**. This applies worldwide

5. *Check-In*

The two pictures of Fig. 3.8 show two extreme choices.

The former (Niguarda Hospital) illustrates a reception service where efficiency is maximized from the hospital's point of view; since the goal is to minimize the time of check-in processes, without particular interest in the emotional condition of patients, they are considered as "numbers" (in fact they are "numbers", having in hand a numbered ticket that identifies their arrival order).

The latter illustrates the reception of a luxury hotel, whose goal is still to maximize efficiency, but prioritizing customer satisfaction: not only guests require a shorter waiting time, but also they should first appreciate the pleasantness of the impact with the organization

We do not argue that the check-in in a hospital should be conducted as in a luxury hotel, but there are intermediate ways, which allow the patient not to feel simply a number, and to be gratified by a relationship of kindness and courtesy, so that she/he can be considered, rather than patient, a customer (as in fact she/he is).[2]

That does not mean that furniture, wayfinding and other aspects are not relevant, but for sure less relevant that the feelings presenting a "promise" of the way of life during the staying there.

---

[2] The check-in at the Humanitas hospital, in Milan, has the appearance, in the procedures and in the way in which one interacts with the customer, of the check-in of a hotel: no splendour in the environment, but kindness and attention by the staff, just like in a hotel. This, as stated by the Communication Manager of the hospital, is the result of a precise and conscious choice by the Management.

Fig. 3.9 Frescoes in the Hospital Santa Maria alla Scala, Siena (Domenico di Bartolo, 1440–1444). The construction of the building (left) and the life inside, with doctors and sick (right). *Source* By Authors

Further examples will be given in Sect. 5.2.

6. *Treatment*

The different types of treatment suggest different intervention mechanisms. However, one element must always be present: the awareness of what is being done and why. As is also evident from Table 3.1, the considerations already made at point 3. Exams can be repeated.

Only out of curiosity, we report in Fig. 3.9 some images of frescoes from the Hospital of Santa Maria alla Scala in Siena (the frescoes were painted around the year 1441, but the building is a few centuries older): despite the fact that at that time the hospital had the function of hosting rather than treating, the presence of doctors was continuous, and the frescoes that decorate the walls of the structure reproduce, in addition to the history of the building (to the glory of the generous donors of the building), also the usual scenes of life inside it, almost as a didactic teaching of the way in which the therapeutic path takes place, in terms of treatment and in terms of life; it is a form of communication towards the awareness of the patient.

7. *Life*

The patient's stay in hospital is in fact a long wait; studded with important moments from the medical point of view, it is for the most part only "time to pass". Free time is not involved in visits and treatments at all, and it is the designer's job (in a broad sense: perceptual aspects and suggestion of activities) to ensure an environment stimulating positive emotions.

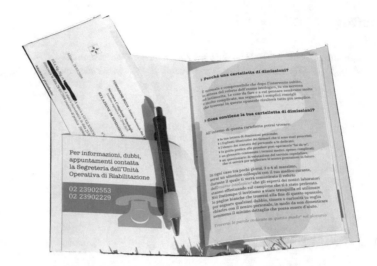

Fig. 3.10 A "Hospital Discharge Passport" was proposed at the Istituto dei Tumori di Milano. Beside the formal discharge letter, it contains instructions to be followed (drugs, checks, food, behaviours, etc.), and a personal diary, to record events, measures and so on. This *passport* contributes to keep the relationship with the Health Care centre, increasing trust in it, increasing *Care* and reducing *Panic-Grief*

We propose various experiences and projects for the life in a Health Care centre in Sect. 5.5.

8. *Visit*

Periodic visits are normal in the course of the *hospitalization*. We can apply the considerations done in point 2.

9. *Check-Out*

There are diseases that, after recovery, have no reason to relapse: for example, infectious diseases or traumas from accidents. There are many other diseases that need to be kept under control, requiring constant medical monitoring: for example, cancer or diabetes. While for the former the discharge from a hospital corresponds to a "farewell", for the second, it is necessary to maintain constant contact with the treatment centre.

Since the patients' expectation leads them to forget the experience and to consider themselves definitively cured, it is advisable to ensure continuous contact with the medical context, without any charge for the patient's psychological state.

So, not only the physical environment of the discharge is relevant, but also the awareness of the patients, together with the capability to keep the relationship with them.

The layout and furniture of the doctor's office (see point 2 above) are relevant to increase *Seeking* and *Care*, as well as the awareness about the disease in any of its aspects (as described above); but is more and more relevant the relationship with doctors and nurses, enabling the continuity of contacts (Fig. 3.10).

References

Defense Acquisition University. (2021). *Systems Engineering Fundamentals.*
Maiocchi, M. (Ed.). (2007). *La comunicazione emozionale nell'ambiente ospedaliero.* Maggioli.

Chapter 4
Measuring Emotions: Kansei Engineering (KE) and Flow KE

4.1 Visceral, Behavioural and Reflective Emotion

In his book *Emotional Design*, Donald Norman brings three levels of human interaction processes: *visceral*, *behavioural* and reflective (Norman, 2013).

Visceral is related to our first impression about something we feel with our senses, which cause the visceral emotional reactions, such as, "it's so cute". Visceral design pays attention to principles like brightness or highly saturated colour (Norman, 2004). Visceral design stimulates the *Seeking* emotion that provides a quick answer (Norman recalls an experience with a 1961 Jaguar E-type as viscerally exciting—see Fig. 4.1). The first glance of a hospital provides information about cleanness, direction, organization and professionalism, spurring a visceral response to a level of competency. For example, an emergency sign arouses this emotion followed by a need for *Care*. Being attracted to an object we immediately recognize, couples the feelings through a quick process between the two emotions, *Seeking* and *Care*.

Behavioural is the human response, which follows the visceral process. Behavioural design considers the performance of an objects function, understandability, usability and physical properties (Norman, 2004). Behavioural design in hospitals can be related to the functionality of a service, object or more. The joy of a good guide designs a service to make their treatment clearer, or the functionality of an appropriate door handle can increase Care and Joy. When analyzing the behaviour of a system the Structure, Function and Material (SFM) model (Shafieyoun & Derksen, 2019) treats these three characteristics as object behaviours as well as human. The structure often described as the context an object or system operates is equally important to the objects function. To use Normans example of the potato peeler that needs to function above all else, neglects the context of having a stove to cook potatoes. The peeler's function becomes irrelevant without the context because the human behaviour is motivated by eating, not peeling.

Reflective processes follow visceral and behavioural levels assessing our feelings affirming our understanding and functionality of the product or the service. The reflective level emotion is dependent on visceral and behavioural level changes. It

© The Author(s), under exclusive license to Springer Nature Switzerland AG 2022
M. M. Maiocchi and Z. Shafieyoun, Emotional Design and the Healthcare Environment,
Research for Development, https://doi.org/10.1007/978-3-030-99846-2_4

Fig. 4.1 Jaguar E-type
(photo from https://www.fli
ckr.com/photos/576643
66@N08/8904901009).
Source From Flickr
Attribution 2.0 Generic (CC
BY 2.0)

can best be explained as the feeling you get after riding a roller coaster, of being proud of the accomplishment, according to Norman (2004). In this scenario the visceral level feeling is often *Fear* and thrill-seeking, which changes to behavioural levels of joy/*Play* emotions (Norman, 2004). More precisely, reflective is the effect of communicating the experience about the other levels. Reflective process for example in the hospital, the entire user experiences are made up of small emotional expressions collected after each step of their journey. The whole experience is carried with them outside of the hospital. As one would expect, a good user experience at each step of the patient journey can have a positive effect on the reflective process.

Still, it is a complex set of emotions that can occur during a hospital visit. The patient may have to deal with negative information. In this study, we have sampled a number of patients at different stages of their care. Participants included people that were required to make choices under difficult circumstances. The reflective process is informed by the visceral information that is not only found in the designed environment. We must be ever mindful of the emotional impact of conditions outside of the designed experience, and yet design has a role to play making sure not to compound a negative experience. These three levels can be mapped to design for the hospital in this way:

Visceral design ➜ Appearance of the hospital;

Behavioural design ➜ The pleasure effective of object, services and experience;

Reflective design ➜ Self-image, personal satisfaction and memories (good user experience).

4.2   Measuring Emotions

Measuring the emotions of people during each part of their journey and understanding their feeling at each level of the process are necessary to address their needs. There are a number of different ways to measure emotions that can be categorized into two groups: *self-reporting measures of emotion* and *automatic emotional measures*.

Self-reporting measures are methods that include surveys, interviews, questionnaires that are taken during or immediately following the activity. Automatic measures emotions using electrodermal, cardiovascular responses, functional magnetic resonance imaging (fMRI) etc. These biometrics measure (blood pressure, skin conductance levels, heart rate and brain activity, eye movement, facial expressions, muscle tone, sweating and vocalizations) can be mapped to emotional response characteristics (Mauss & Robinson, 2009). It was important to examine the component parts of interior spaces and their individual effect on the emotional state of the participants in this study. Using automatic measures could only provide a holistic view of the emotional effect which could not discern between the parts of a space, and for these reasons self-reporting methods were more appropriate.

4.3 Kansei and Kansei Engineering

Among the methods of both self-reporting and automatic, Kansei Engineering (KE) is used to measure participant's emotions to design effective products. Before introducing the KE method, a note on the origin of the word Kansei is in order:

kansei-kai 感性界 (哲) can be interpreted as sensible, in Latin, "mundus sensibilis" and the German translation of kansei-ron 感性論 (哲) is the aesthetic, "Ästhetik".

According to Mitsuhiko Toho (2006) who provides a detailed semiology of the term provides these four features:

1. Receptivity of sensation or perception stimulated in the external world through sense organs;
2. Feeling, impulse or desire stimulated by sensation;
3. Sensuous desire which should be controlled by rational nature or *animus quo*;
4. Sensuous cognition as a material of thinking.

More accurately in this context; being a higher function of the brain (Harada, 1998) popularized in the theory of sensitive cognition systematized by Baumgarten as the aesthetics. In Kant's works, *Ästhetik* it is a theory of sensuous intuition which forms the foundation of the intellectual (Toho, 2006). In modern science, the psychological phase of Kansei is an outcome through cognition and the five senses. The Fig. 4.2 shows the process of Kansei and the five senses within the structure of the brain (Lokman & Kamaruddin, 2010).

KE is a technology that integrates Kansei (feelings and emotions) with engineering methodologies. It is a field in which the development of products that deliver happiness and satisfaction to users is conducted technologically, by analyzing human emotions and incorporating them into product design (Nagamachi & Lokman, 2009) (Fig. 4.3).

Kansei Engineering is not only an evaluation method at the end of a design project; many companies used KE throughout the design stages of products including Mazda, in the development of the Miata model, Wacoal designs underwear and Sharp to

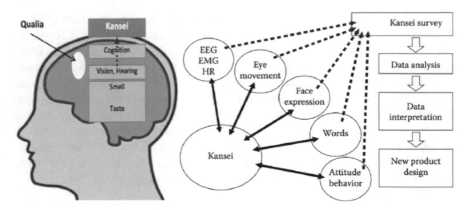

Fig. 4.2 The process of Kansei (Nagamachi & Lokman, 2009). *Source* By Authors

Fig. 4.3 The Miata model of Mazda, examples of Wacoal underwear (photo Tony Tseng), the Sharp camcorder. *Source* From https://it.wikipedia.org/wiki/Mazda_MX-5#/media/ File:1st_Mazda_Miata.jpg–publicdomain; https://commons.wikimedia.org/wiki/File:Wacoal-304 15340034.jpg-CreativeCommonsAttribution2.0Generic

produce VTR camera instead of a conventional ocular lens (Schütte et al., 2004). KE helped these companies to increase their sales by designing something people had emotional attachment. KE is a way to design products which bring happiness and satisfaction to people bound in loyalty to these companies. Nagamachi in 1970, noticed that good quality products are preferred by customers. The good quality products are not just the products which last longer, they are the products people love and desire them because of qualities outside of features or function. In an effort to uncover this relationship, Nagamachi invented KE to discover and respond to the feelings people have towards products. Initially KE was used to examine component parts of a project, separating the different elements for analysis, identifying the Kansei of each element in each product. Understanding the emotional response people have about each segmented element of a product such as shape, colour, material then assemble the highest ranking elements together to design the final product (Nagamachi & Lokman, 2011).

Since 1970 when Kansei engineering started at Hiroshima University, eight types of Kansei Engineering have been developed as techniques to evaluate products in different segments of the market. KE type I: Category classification, KE type II: KE

Table 4.1 Kansei engineering types

KE Type	Subject
I	Category classification
II	KE system
III	KE Modelling
IV	Hybrid KE
V	Virtual KE
VI	Collaborative KE
VII	Concurrent KE
VIII	Rough Set KE

system, KE type III: KE Modelling, KE type IV: Hybrid KE, KE type V: Virtual KE, KE type VI: Collaborative KE, KE Type VII: Concurrent KE, KE, Type VIII: Rough Set KE (Ishihara et al., 2005; Nagamachi, 2003).

Since 1970 when Kansei engineering started at Hiroshima University, eight types of Kansei Engineering have been developed as techniques to evaluate products in different segments of the market (Ishihara et al., 2005; Nagamachi, 2003) (Table 4.1).

4.3.1 Design Waiting Area Using Kansei Engineering Type I

In order to explain the method used in applying Kansei Engineering in our Emotional Design proposal, we describe a project we carried on studying the hospital Waiting Areas. This part will be further recalled in the part of the book in which we will examine possible actions for the Waiting Areas (Chapter 5).

Few studies used KE to design services and experience in the process. Inspired by the work of Dahlgaard (Dahlgaard et al., 2007) who uncovered the value of KE, we used Kansei engineering type I to understand the people emotions experienced while waiting for health care services. Using the seven fundamental emotions defined by Jaak Panksepp, we measured levels of *Fear, Rage, Grief, Seeking, Care, Play* and *Lust* (Panksepp & Biven, 2012) on 200 cancer patients. We used the Kansei Engineering method to analyze survey data from four waiting areas in two different hospitals in Milan, Italy. Complementing the survey data, we collected information using a number of well-established design tools: patient journeys, scenarios, storyboards, personas and interviews. Each data set were analyzed using appropriate methods such as semantic differential scales for the questionnaires Snider & Osgood, 1969), and factor analysis to examine data generated by participant interviews. Analysis results demonstrated how different design elements and social characteristics of waiting areas could influence the emotional experience of their users. For example, we verified that warm colours, abundance of light, curvilinear furniture layout and the presence of art works could enhance patients' positive emotions. Additionally, their state affected perception of their experience and, ultimately, their perceived quality of care.

The study of "Environmental Effect on Emotion in Waiting Areas Based on Kansei Engineering and Affective Neuroscience" (Shafieyoun et al., 2014) selected KE type I to breakdown the waiting areas to different segments and discover it from user's emotion of view. This study would offer the characteristic of a new waiting area based on peoples' emotion. KE type I is an iterative process (Fig. 4.4) and includes: choosing a domain, two parallel steps "span of the semantic space" and "span the space of property", synthesis the result, test the validity and repeat the process until it is suitable for implementation (Fig. 4.5).

Choosing domain: Choosing the target group is the first step of the process. In our study, we chose two hundred patients who are diagnosed with cancer at different

Fig. 4.4 KE model proposed by Schütte (2005). *Source* By Authors (elaborated from a catalog picture)

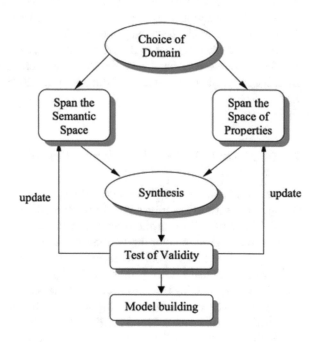

Fig. 4.5 Age distribution among participants. *Source* By Authors

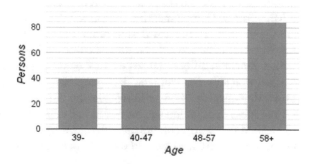

stages of their treatments. Fifty patients for each four waiting areas (WAs) were distinguished in two hospitals (Istituto Nazionale dei Tumori and Besta in Milan, Italy). The range of the patients were above 40 and the majority more than 58.

Spanning the semantic of Space: In this step, we collect Kansei Words (KWs) to describe the environment. The sources for the KWs can be magazine, literature, conversations, experts, internet and so on. KWs can be between 50 and 600 words (Nagamachi, 1997). KWs clustering manually or statistically (Schütte, 2005) allows to reduce them to 15–25 categories. In our study, we use KWs to describe the environment of the WAs by obtaining 650 KWs. We collected them from magazines, internet, books, papers and dictionaries, as well as from personal observations. In the next step, we used manual expert method to perform the word structuring. We grouped 325 of them by omitting the synonyms and grouped them together. Finally, reduced them to 24 KWs, classified into 6 out of 7 primary-process emotions. We excluded LUST as it is difficult to apply to the hospital environment.

Spanning the space of properties: In this step, we collect few samples of the excising, products in our case are some samples of WAs. We chose four different waiting areas in Istituto Nazionale dei Tumori (Figs. 6.1 and 6.3) and Besta hospital (Figs. 6.2 and 6.4) in Milan, Italy. WA number is the WA of breast radiology department at Istituto Nazionale dei Tumori, the design process described in Chapter 1 (Fig. 4.6).

Synthesis: In this step, Semantic Space and the Space of Properties link together. For each KWs, there are some aspects of the product to affect the KW. People relate each KWs to an emotion, in this part we understand the people's definition of each KW. In our study, we created a 7-point semantic differential method (SD) questionnaire based on KWs, ranging 1–3 (*not so good*), 4 (*neutral*) and 4–7 (*very good*). We asked participants to answer it based on their feeling about colour, material

Fig. 4.6 Istituto Nazionale dei Tumori and Besta in Milan, Italy. *Source* By Authors

Fig. 4.7   7-point semantic differential method questionnaire. *Source* By Authors

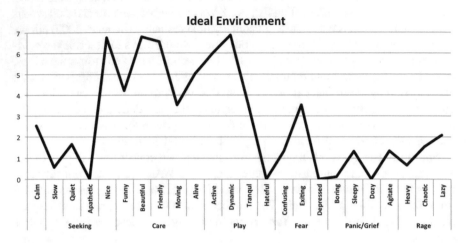

Fig. 4.8   An ideal waiting area seen by designers. *Source* By Authors

and furniture layout of the WA. Fifty participants in each WA answered the questions (Fig. 4.7).

Before analysis participants' respond we asked ten designers to use the same KWs and tell us about the ideal WAs based on the user experience, the result showed the KWs "Beautiful", "Friendly", "Nice", "Active", "Alive" and "Dynamic" have higher rates (Fig. 4.8).

The data were treated with factor analysis, with the purpose of semantic structuring. This was done in SPSS (Statistical Package for Social Sciences), a commercial statistical software. The input data were the mean values for the 24 KW's describing each of the 4 WA's and including an ideal WA. We used this method as the factor analysis to simplify this raw data and to reveal connections between each of the WA's. Results below demonstrate new categories of KW by performed factorial analysis (Table 4.2).

All KWs were reduced to five factor-words: *Gloomy, Friendly, Active, Confusing* and *Agitated*, which had the highest loading in more than one analysis. Factor 1 (*Gloomy*) for colour is: "*Tranquil*", "*Calm*", "*Quiet*", "*Boring*", "*Sleepy*", "*Dozy*" and "*Depressed*". We can observe that there are no adjectives +4 in Ideal WA. Factor 2 (*Friendly*): "*Nice*", "*Funny*", "*Beautiful*" and "*Friendly*" are +4 in ideal WA. According to the amount of Kansei Rating in 4 WA we can observe that WA 1 has +4 rating in these adjectives. Factor 3 (*Active*) "*Alive*", "*Active*" and "*Dynamic*" we observe +4 in Ideal WA and WA 1 having all these adjectives +4. Factors 4 and 5 respect the similar trend. Finally, we find out that the colour of WA 1 has

Table 4.2 Kansei words per Colour

Factor 1: Gloomy	Factor 2 Friendly	Factor 3 Active	Factor 4 Confusing	Factor 5 Agitated
Tranquil	*Nice	Moving	Confusing	Agitated
Calm	*Funny	*Alive	Exciting	Heavy
Quiet	*Beautiful	*Active		Chaotic
Boring	*Friendly	*Dynamic		Hateful
Sleepy	Slow			Lazy
Dozy	Apathetic			
Depressed				
	WA1	WA2		

*Adjectives 4+ are rating for Ideal Waiting Area

been perceived friendlier and has more positive emotion for users. Results below demonstrate a comparison of an average participants of 1–7 ranged values among all 3 attributes: colour, furniture layout and material.

Conversely, Factor 1 in Table 4.3 is "*Slow*" and includes "*Hateful*", "*Tranquil*", "*Quiet*", "*Slow*" and "*Apathetic*" which has no +4 adjectives however having instead 3 adjectives ("*Hateful*" and "*Apathetic*" and "*Slow*"), which are less than 1 in Ideal WA. We can also observe that the WA's 2 and 3 have +4 rating of Kansei in these adjectives. In addition, in Table 4.3 we observe the results of negative emotions in WA's. Table 4.3 shows WA's 1 and 3 as favourite selection for furniture layout.

Under the word "Material" we consider any material in environment, which can be tangible material in furniture or a wall texture. We can see in Table 4.4 that users chose WA1 as the best WA in case of material.

We believe that decreasing negative emotion and increasing positive ones in new WA's would be possible by knowing better the perception of people. Though, we

Table 4.3 Kansei words per furniture layout

Factor 1: Slow	Factor 2 Nice	Factor 3 Active	Factor 4 Dynamic	Factor 5 Chaotic
Hateful	*Nice	Moving	*Dynamic	Agitated
Tranquil	*Beautiful	*Alive	Confusing	Heavy
Quiet	*Friendly	*Active	Exciting	Chaotic
Slow	Sleepy			Calm
Apathetic	Dozy			Depressed
	Lazy			*Funny
				Boring
	WA1, WA3	WA1. WA3	WA3	WA3

*Adjectives 4+ are rating for Ideal Waiting Area

Table 4.4 Kansei words per material

Factor 1: Sleepy	Factor 2 Calm	Factor 3 Active	Factor 4 Confusing	Factor 5 Agitated
Slow	*Beautiful	Alive	Hateful	Agitated
Apathetic	*Friendly	*Active	Confusing	Heavy
Boring	Depressed	Dynamic	Exciting	Chaotic
Sleepy	Tranquil	*Nice	Moving	Dozy
	Calm	Funny		Lazy
	Quiet			
	WA1	WA1		

*Adjectives 4+ are rating for Ideal Waiting Area

Table 4.5 Kansei words with 7 Panksepp's Emotions per WAs

	Positive		Negative		
Emotion	Care	Play	Fear	Seeking/Grief	Rage
Colour					
KW	Friendly	Active	Gloomy	Confusing	Agitated
WA	WA1	WA2	WA3	WA2, WA3	WA4
Furniture Layout					
KW	Nice Funny	Active Dynamic	Hateful	Slow	Heavy
WA	WA1, WA3	WA1, WA3	WA3	WA2	WA4
Material					
KW	Calm	Active	Agitated	Apathetic	Boring
WA	WA1	WA1	WA3	WA2,3	WA2,3

can depict negative elements of each WA by choosing adjective less than 1 in Ideal WA and +4 in our WAs. The result is shown in the Table 4.4. We merged *Seeking* with *Grief*. As we know, *Seeking* is considered as *the mother emotion* (Panksepp et al., 1998), which serves as a channel to enhance other ones. Results below depict comparison of Kansei Words of four WAs with negative and positive adjectives for three characteristics: colour, furniture layout and material. In overall, WA1 has more positive emotions for all three factors. WA3 has a friendly Furniture Layout but other factors dominate negative emotions. By analyzing the character of WAs, we obtained a better insight in perception of participants (Table 4.5).

4.3.2 Flow Kansei Engineering

The available information helps us to build a new Ideal Waiting Room from the designer point of view, based on patients' feelings. This research did not, however, take into account the state of mind patients before they entered the waiting areas. We believe people might be affected by different emotions before arriving at the hospital, and it might influence their feelings about the environment and services as well, as a follow-up, we added *flow* (Csikszentmihalyi, 1992), a subconscious state of mind to create Flow Kansei Engineering (FKE). The FKE method encourages people to enter a flow state, before responding to questionnaires related to their emotions in waiting areas. Flow testing measured their emotional state at three different phases of distractions or engagement with their treatment. Investigations into the effect's art have on distracting patients in waiting areas were also done at two hospitals, in Milan (Istituto Nazionale dei Tumori) and in Rome (San Camillo). The results of the FKE study indicate that actions such as engaging people in a conversation about their treatment are useful distractions that raise positive emotions. A conceptual framework was proposed as a model to follow, and consistently perform a trustable and comparable method centred on the patient's perception of hospitals, as a form of user experience (UX) study. This method was tested with small numbers of participants; however, it has the potential to become a guideline for every waiting area and waiting time (Fig. 4.9).

In FKE in first step **choice of domain,** we consider the participants habits and interests to be able to enter people in their flow mode. In **Spanning the Semantic Space** in additional to gathering KWs, we gather words related to Flow as well. In **Span of the space of properties**, we remember to look at the product and see if they have a potential to help people to enter to their Flow mode, for example, paintings in waiting areas might be one of the opportunities. In **synthesis** before distributing the questionnaires we enter participants to their Flow mode and let them to stay in it for a while then (about 20 min). Then, we ask them to fill the questionnaires and tell us how they feel about the colour, material and furniture layout of the WA.

4.3.3 Flow Test

The following we proposed a Flow test to validate our method by measuring the people's emotion with the same SD questionnaire in three different created situations in WA.

- In an empty WA with no distractions, empty walls, no music, no TV; ...
- Engage people to a conversation related to their treatment, play a video about their treatment journey, a story from a volunteer cancer survivor;
- Preparing a useful distraction like paintings, Sculpture, music (Depends on the culture, in Italy might be Opera) (Figs. 4.10, 4.11, 4.12).

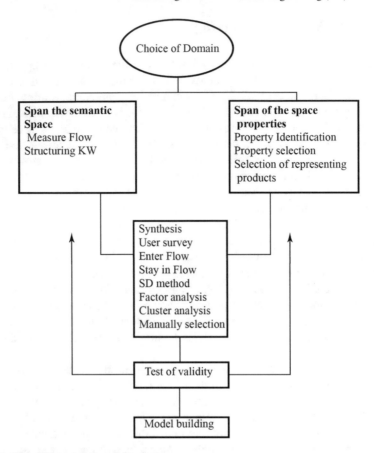

Fig. 4.9 Flow KE model. *Source* By Authors

Fig. 4.10 Waiting area with no useful distraction. *Source* By Authors

Fig. 4.11 Waiting area with useful distraction (Paintings). *Source* By Authors

Fig. 4.12 Waiting area with useful distraction (Educate patient). *Source* By Authors

The result of the test shows how much people need *Care*, engaging people in the conversation related to their treatment and creating useful distraction gave more *care* to people than just leave them with no distraction and let them to find their own way to spend their waiting time. FKE's effect on visceral, behavioural and reflective level is deniable and can create a good user experience.

4.3.4   Painting as a Useful Distraction in Waiting Area

Useful distractions can effect on decreasing the perception of time and increasing positive emotion specially *care*. Distraction is used for pain control as well (Gijsbers & Melzack, 1967). According to the result related to the Flow test, we followed a special Exposition of Artworks, named "When Art meets Care" in the main WA of Oncological Centre INT, and another one at the Istituto dei Tumori (Milan, Italy), introducing the paintings of a Japanese painter, Tetsuro Shimizu. He drafted the artworks of his project "Immunity" during his cancer treatment. The patient's emotion was measured during the exhibition and after that (Figs. 4.13 and 4.14).

Fig. 4.13 Painter (Tetsuro
Shimizu) and one of his
painting. *Source* By Authors

Fig. 4.14 A main Waiting
Room in Instituto Nazionale
dei Tumori—Milan, Italy
during an exhibition. *Source*
By Authors

In this study we measure the emotion of people with and SD questionnaire in two phases, during and after the exhibition. 200 participants during the exhibition and 200 after the exhibition. In the first stage, our study subjects were around 200 participants which presented in three days main waiting room of INT. Between 200 questionnaire which were divided 153 questionnaires were valid. In the second phase among 200 questionnaires, 165 were valid. The first phase the majority of participants are more than 58 years old but in second phase population are younger. The same as other studies above here we categorized emotions based on seven emotions of Panksepp (Panksepp & Biven, 2012) except *Lust* (Fig. 4.15).

The results show during the exhibition *Play* and *Care* increased, people were **feeling** that the environment is more active and friendly and the same time more agitated which we can refer it to the expressionism characteristic of the paintings. We repeated the same survey with the same questionnaire in a WA for cancer patients in San Camillo Hospital in Rome. 200 participants answered the questions during and after the exhibition. The majority of participants are between 48 and 57 years old during the exhibition and after the exhibition the age distribution was between 39 and 57. The results show people's feeling does not change during and after the exhibition. Considering the enhancement of monotone mood during exhibition we can refer it to the place of the painting installations. Painting was installed more on the way of people to the WA and people were not interacting with them during their waiting time (Figs.4.16, 4.17 and 4.18).

4.4 Measuring the Emotion of Patients During Their Journey in the Hospital

These studies focused on both waiting time, waiting areas in health care centres. Interactions in waiting areas were studied by knowing patients feeling in different waiting areas with different colour, material and furniture layout and effect of visual arts on their emotion. Waiting time was examined by user experience and causes of long wait based on observation and literature reviews.

According to Karapanos, user experience changes over the time and we can describe temporality of experience in three phases: Orientation, Incorporation to identification (Karapanos et al., 2009). Erik Stolterman proposed an alternative way to study and describe temporal pattern of interaction (Stolterman & Jung, 2011).

Record patient activities can happen, with an application, questionnaire, observation, videography and interviews. Knowing about their age, sex, lifestyle and their

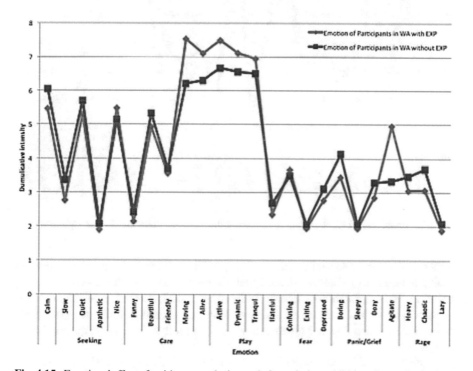

Fig. 4.15 Emotional effect of waiting room during and after painting exhibition. *Source* By Authors

Fig. 4.16 Painting Installation in waiting area in San Camilo Hospital in Rome-Italy. *Source* By Authors

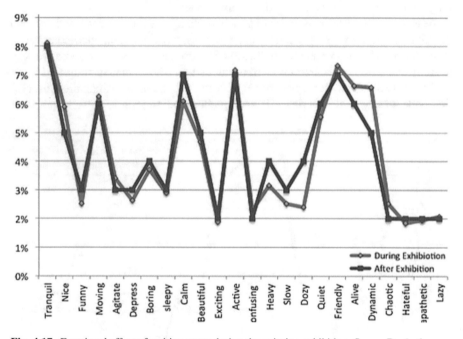

Fig. 4.17 Emotional effect of waiting room during the painting exhibition. *Source* By Authors

perception of time and real duration of time will be helpful. All results should be coding to build the verbal and visual vocabularies based on following groups: focus, attention, actions, intention of being in hospital and element of time. Then, every group will be divided to orientation, Incorporation to identification. Then after analysis of the obtained data to implementation in design then we can verify it to better design.

Finding interaction in waiting time in hospitals needs record activities in every short moments of user experience in whole patient journey. Events, activities, actions,

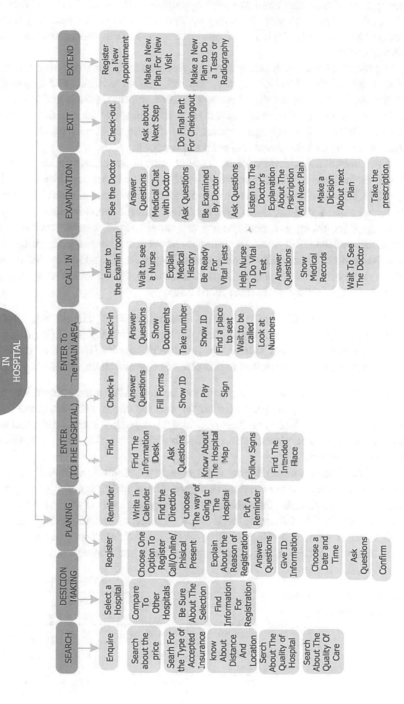

Fig. 4.18 User experience in patient's journey in the Hospital. *Source* By Authors

perception of time and emotions are happening in the same time and they will be changed in time sequence. Focusing on specific moment of patient journey will provide better result. Dividing patient journey to several short journeys makes the survey easier.

Elements of time such as duration, session and frequently of every act should be discovered in three level of paying attention, getting involved and interacted with environment. Finding relation between patient emotion, action and the difference between perception of time and actual time will show the amount of patient's satisfaction. Beside all above age, gender and lifestyle of users have a role on their interaction with environment and even in perception of time.

Figuring out the attention, focus, element of time, intention of act and notification based on the time in whole journey will be helpful to know user experience better.

Observation is the first choice to people and record them and obviously it cannot be accurate. Storytelling, flow chart, sketches, graphic scenario, speaking, diagram, writing, videography and photography are other helpful ways to review event and activities, and the result usually is summarized event and some special minor activities will be ignored. Thanks Erik Stolterman to introduce Visual and verbal component in his survey in 2011 (Huang & Stolterman, 2011). Based on his work below I will propose a preliminary method for this study:

- To create a questionnaire to know the time sequence of patient's act, emotion and their perception of time during their journey in the hospital. It helps us to know more details in patient journey by keeping continuity. The use of the model of patient experience presented in this chapter can enrich the study.

Patient can fill the forms using their own language. After the completion, we categorize acts and emotions (in particular we use the seven basic emotions following Panksepp). Acts can be categoried to pay attention, get involved and interact or orientation, incorporation, identification groups based on duration session and frequency of time (Table 4.6).

This method was tested in a small group and the result showed *Seeking* and, *Fear* in arrival and enter to the hospital is higher. *Grief* in Engage stage is a part of the journey. We are planning the validity of the proposed method with a larger sample of participants. We are following the journey of selected patients through the hospital system to examine the effect of long waits on patient experience and expressed emotional states. It is our hope that future research may show a correlation to patient

Table 4.6 Sample of form filling questionnaire

Emotion	Action	Time perception	Actual time
Anxiety	Waiting sitting	20 min	6 min
Attention to a sound	Standing	1 min	5 min
Delusion	Sitting again	20 min	5 min
…	…	…	…

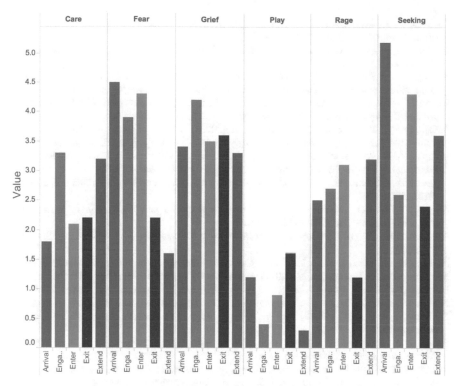

Fig. 4.19 An example of measures on a small group of patients; the "quantity" of the various emotions perceived along the various steps in the waiting rooms. *Source* By Authors

recovery times. Waiting time and the character of waiting provide people with the opportunity to be by themselves, to take a moment to contemplate, to think and to make decisions. According to Dabrowski, people can develop their personality by taking advantage of the hard moments in their life (Dabrowski, 1967) just by thinking through the situation. A question that has motivated some of our research recently is, how can design be used to enrich these moments of contemplation (Fig. 4.19).

The method was tested with a small group of people. It shows in patients' journey.

References

Csikszentmihalyi, M. (1992). Flow: The psychology of optimal experience. *Journal of Leisure Research, 24*(1), 9394.

Dabrowski, K. (1967). *Personality-shaping through positive disintegration*. Little, Brown.

Dahlgaard, J. J., Khanji, G., & Kristensen, K. (2007). *Fundamentals of total quality management*. Routledge.

Derksen, G., & Shafieyoun, Z. (2019). Prototyping a new economy. *LearnXDesign Conference*. https://doi.org/10.21606/learnxdesign.2019.14119

Gijsbers, K. J., & Melzack, R. (1967). Oxygen tension changes evoked in the brain by visual stimulation. *Science, 156*(3780), 1392–1393.

Harada, A. (1998). On the definition of Kansei. *Modeling the Evaluation Structure of Kansei 1998 Conference, 2,* 22.

Huang, C-C., & Stolterman, E. (2011). Temporality in interaction design. *Conference: Proceedings of the 2011 Conference on Designing Pleasurable Products and Interfaces.*

Ishihara, I., Nishino, T., Matsubara, Y., Tsuchiya, T., Kanda, F., & Inoue, K. (2005). *Kansei and product development* (Ed. M. Nagamachi. Vol. 1.) (In Japanese) Kaibund.

Karapanos, E., Zimmerman, J., & Martens, J. (2009). User experience over time: An initial framework. *Proceedings of the SIGCHI Conference on Human Factors in Computing Systems.*

Lokman, A. M., & Kamaruddin, A. (2010). Kansei affinity cluster for affective product design. *International Conference on User Science and Engineering (i-user),* IEEE.

Mauss, I. B., & Robinson, M. D. (2009). Measures of emotion: A review. *Cognition and Emotion, 23*(2), 209–237.

Nagamachi, M. (1997). *Kansei engineering: The framework and methods,* Kansei engineering 1(M. Nagamachi). Kure, Kaibundo Publishing Co.

Nagamachi, M. (2003). *The story of Kansei engineering* (in Japanese) (Vol. 6). Japanese Standards Association.

Nagamachi, M., & Lokman, A. M. (2009). *Kansei engineering: A beginners perspective.* KDRU.

Nagamachi, M., & Lokman, A. M. (2011). *Innovations of Kansei engineering.* CRC Press.

Norman, D. (2004). *Emotional design: Why we love (or hate) everyday things.* Basic Books.

Norman, D. (2013). *The design of everyday things* (Revised and expanded ed.). Basic Books.

Panksepp, J., & Biven, L. (2012). *The archaeology of mind: Neuroevolutionary origins of human emotions.* W. W. Norton Co.

Panksepp, J., Knutson, B., & Pruitt, D. L. (1998). Toward a neuroscience of emotion. In *What develops in emotional development?* Springer.

Schütte, S. T. W. (2005). *Engineering emotional values in product design: Kansei engineering in development.* PhD diss., Institutionen för konstruktions-och produktionsteknik.

Schütte, S. T. W., Eklund, J., Axelsson, J. R. C., & Nagamachi, M. (2004). Concepts, methods and tools in Kansei engineering. *Theoretical Issues in Ergonomics Science, 5*(3), 214–231.

Shafieyoun, Z., Radeta, M., & Maiocchi, M. (2014). Environmental effect on emotion in waiting areas based on Kansei engineering and affective neuroscience. *International Conference on Kansei Engineering and Emotion Research.* Linköping.

Snider, J. G., & Osgood, C. E. (Eds.). (1969). *Semantic differential technique: A sourcebook.* Aldine.

Stolterman, E., & Jung, H. (2011). Form and materiality in interaction design: A new approach to HCI. *Proceedings of the 2011 Annual Conference Extended Abstracts on Human Factors in Computing Systems.*

Toho, M. (2006). *What is Kansei* (Unpublished paper presented at Interfejs użytkownika - Kansei w praktyce Conference), Warszawa.

Chapter 5
Experiences

5.1 Some Forewords

A scientific approach should be carried on starting with some hypothesis to be validated with experiences and measures (typically with the described Kansei Engineering). Sometimes an experience comes before the ideas and suggests some hypothesis to be verified, and sometimes, the validated models drive the experiences and the projects.

The experience described here in some cases is just the consequence of validated models, and sometimes the starting point of a new model. In the following descriptions, we will not distinguish between the two types.

Many of them are the result of workshops carried on with students of the last two years of the Politecnico di Milano, Faculty of Design: in this case, all the authors are referred.

Details of many of those experiences can be found in various books (Maiocchi, 2007; Maiocchi, 2008; Maiocchi, 2010).

5.2 Registration at Istituto Nazionale dei Tumori[1]

The Istituto Nazionale dei Tumori in Milan is a primary cancer Health Care centre, established in the first decades of the last century, and now involving more than 2,000 employees, and providing more than a million of medical services (visits, analysis and exams, hospitalizations, etc.) every year.

The entrance from the street leads to a huge atrium full of services and information: a reception, a help desk, offices of the Unit for the relations with the customers,

---

[1] The experience is related to an educational workshop held at Politecnico di Milano, for a proposal for the Istituto dei Tumori in Milan. The project had contributions by Ambra Farris, Emanuele Mantovani, Elisa Montalbano, Flavia Pellegrinelli. A detailed description can be found in (Maiocchi, 2010).

various corners in which many associations present supporting services (voluntary associations for psychological support, or providing any kind of services to the patients and to their families), a dense official signage for the medical services, information about conferences and exhibitions within the centre, information about events out from the centre, info points, monitors communicating various kind of information, private advertising, small waiting areas with seats and so on.

The continuous growth of both the centre and the number related activities leads to an overlapping of too much information, not always updated. Visitors are disoriented and their *Seeking* activity suffers.

Moreover, the visitor cannot understand which services are available, and where: for example, while the reception is easily identified immediately after the entrance, the help desk is hidden by columns and panels; but the former works mainly as concierge for the personnel and it is not clear to the visitors. Again, there is not a uniform style in furniture, in colours, in information formats and so on, as shown in Fig. 5.1.

The re-design of the atrium, according to the considerations done in Chap. 3, was organized as follows (Fig. 5.2):

Fig. 5.1 Disorienting organization in the atrium of Istituto Nazionale dei Tumori. *Source* By Authors

Fig. 5.2 The new layout proposed for the atrium (entrance lower left corner). *Source* By Authors

The entrance leads directly to a conceptual map, with different colours code, and with various paths conducting to the various indicated services: this should improve the *Seeking* activity (Fig. 5.3).

Two red blocks are easily identified: on the right after the entrance, for information and on the left, centred, for all the voluntary associations.

The former, groups both concierge and help desk, with a concave counter (recalling a hug, a welcome, enhancing *Care*) (Fig. 5.4).

The latter cuts out a door in the panel (to express you are entering in a different context) and draws hands to recall offers and solidarity (*Care*) (Fig. 5.5).

After the entrance, on the left, the small waiting area has bend restructured substituting the metallic chairs with upholstered armchairs (soft, again *Care*), and adding small tables reporting on their surface the information to reach common services (toilets, bar, newspapers and so on) (*Seeking*) (Fig. 5.6).

Fig. 5.3 The central map and the pathways. *Source* By Authors

Fig. 5.4 Concierge and info point. *Source* By Authors

Fig. 5.5 The voluntary associations area. *Source* By Authors

Fig. 5.6 The waiting area at the entrance. *Source* By Authors

In Chap. 3, we introduced the problem of the first contact with the centre through a telephone call or through online procedures.

The use of call centres, in general oriented to reduce costs, rarely is respectful of the customers; it is important:

- to guarantee an answer by the operators within a few seconds;
- to reduce the depth of the question's hierarchy;
- to allow as much as possible and as soon as possible the contact with a human operator.

The user interface of an online service must be properly designed according to the best web design practices, and graphics, paths and contents must be organized according to the principles of Emotional Design.

For example, the interface and the content for an info point for the atrium of the Istituto dei Tumori are based on menus composed of coloured "leaves" (*Play*), populated by cute little puppets scribbled with a child's hand (again *Play*). All the pages are inspired to the same style, and the user can ask for a printout of some pages (for example, a small map with the pathway to reach some offices or places (Fig. 5.7).

The richness of information (from the history to services, to how to spend free time, to the life in the surroundings of the Istituto) not only satisfies, but also stimulates *Seeking*), cover also the awareness about the services, and the status of a procedure: for instance, it is possible to know the average time for getting a medical service) (Fig. 5.8).

The same information is available online.

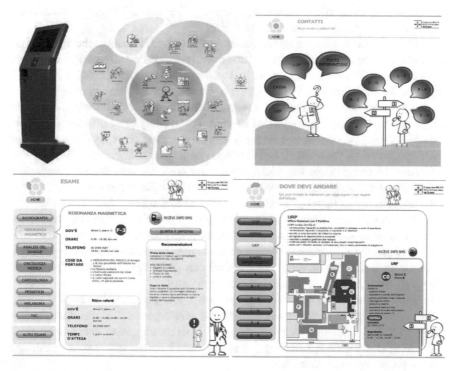

Fig. 5.7 Info-point. The tower, the initial menu, a second level menu, rules for an exam, pathway for a possible printout. *Source* By Authors

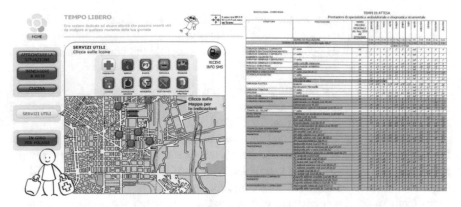

Fig. 5.8 Info-point. What to do in the surrounding of the Istituto, and a table of the average time to wait (how many days), measured for each service in the various months. *Source* By Authors

5.3 The Department of Breast Radiology at the Istituto dei Tumori of Milan

The Istituto dei Tumori in Milan growth quickly, and temporary solutions to the high requests were often needed. So, the department of Breast Radiology was transferred in a large area in the basement of the building.

After some years, a strong refurbishment was planned; thanks to the suggestions of the responsible of the Office for the customers relationships, Roberto Mazza and his deputy Ivan Pozzati), the director of the department, doctor Silvana Bergonzi, suggested a collaboration with the Design department of Politecnico di Milano.

Due to economic constraints, it was decided to approach the topic through an official educational workshop, so that students could study the context and provide proposals. After the course, the selected solutions should have been implemented under the control of the student, and they had got some rewards.

Despite the many interesting ideas, the constraints were very strong, both for the sanitary standards and for economic limits. So, the only actions proposed were related to the temperature of lights (warm), to colours and to the disposition of furniture.

All the doors were selected with a colour code: deep blue for offices, red for the laboratories, acid green for locker rooms, dark green for passages and so on. The walls were painted with intense colours or were supposed to be filled with paintings.

So, a large waiting room was painted in dark orange, and purple upholstered chairs were positioned in diagonal lines, breaking symmetry.

In order to enrich the walls through paintings and poster, it has been decided, instead of usual print, to ask to some artist to donate a piece. The result was astonishing: over the course of a month, through word of mouth, more than one hundred artists from seven different countries sent over one hundred and seventy paintings and sculptures, often made ad hoc!

Those pieces were hanged on every walls: in the waiting room, in the offices, in the laboratories and in corridors (Figs. 5.9, 5.10, 5.11, and 5.12).

The effects of those simple interventions were impressive: both visitors and patients, doctors and staff changed their behaviours; once entered in the department, they stopped and looked around, as if to understand what was happening. They were surprised that the environment was too different from what they would have expected in a hospital. *Seeking* was growing: the need to grasp the novelty and to understand made them to forget their illness problems. Of course, as mentioned in Chap. 1, the principles of Ramachandran played a relevant role, in particular *contrast, symmetries* and *metaphors*, often present in the paintings. Most important, the *contrast*, not only

Fig. 5.9   A corridor in the Breast Radiology Department. Before the refurbishing (left) and after (right). Note the dark green door, the many paintings and the brilliant lights. *Source* By Authors

Fig. 5.10   The waiting room (already shown). *Source* By Authors

Fig. 5.11 A doctor's office. Few actions, but paintings, were selected so to get heavily coloured spots on the walls. *Source* By Authors

Fig. 5.12 A part of the received paintings, many of them very large. *Source* By Authors

in the colours, but also from a conceptual point of view: the *contrast* between the expected and the actual environment. In Chap. 4, we mentioned this waiting area was selected as the most friendlier, comfortable and warmer among four chosen waiting areas in our analysis and measurement.

Unfortunately, we discovered this effect of the changes too late, and we had not set up some kind of measurements about the medical parameters. We only recognized the growth of the number of smiles among the people entering the area and collecting declarations by some customers (Fig. 5.13).

Italian culture makes them to live in art and appreciate it. Paintings and sculpture got the attention of every patient who were walking to the radiology waiting room. Sometimes they were standing and watching the paintings and some of them were sitting and looking around to discover the environment. The prepared book about their paintings was taking some of their time to learn about each painting. The patients note at the end of their visit showed a lot of positive ideas, excitement and joy during their waiting time. Patients and their family were happy to be in a different place

---

Thanks for the many messages of courage and hope!

To overcome problems, to help others and to love the world, you must first love yourself. Only by loving and respecting yourself can you give yourself to others, and you can be of help ... Nice message!

Reading about art while waiting for an exam helps to forget, not to think why we are here.

Fortunately, there is art! An expression is worth more than it means!

A flight into space! Art is an expression of life, of love, of harmony and disharmony with ourselves and others. Nothing more suitable for this exhibition. Thanks, you were fabulous!

While I'm waiting, I look at a painting. It reminds me of the Universe with its bright colors.

It is a wonderful thing! All waiting rooms should be so welcoming and pleasant!

All beautiful, all different! My eyes light up to see them! I often pass by on purpose, I take pleasure in watching them! There are still so many walls in the hospital to cover with your paintings! I'll come back.

This morning, when I came back for the ultrasound, I counted camels in the xxx picture. Time has passed.

Good morning and thank you.

In particular moments it becomes difficult to appreciate the beauty of a painting. The thoughts are many. Pictures and paintings that refer to "rebirth" have been my favorites.

A touch of color and serenity is important beside organization and cleaning. The walls so covered with paintings are gorgeous! Colors, lots of chlorines. Beyond black and white, good and evil: in this state gray we are all gray. You must think positively and love yourself: a lot, a lot! I think the painting is the revelation of the soul of every artist. Thank you everybody!

Fig. 5.13 Some of the comments left by visitors. *Source* By Authors

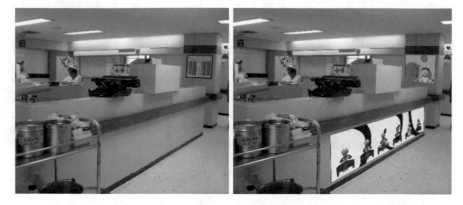

Fig. 5.14 Paediatric Reception—Hospital 33—Queretaro (Mexico), Stories explaining the parcourses in the hospital have been added on the desk (Catalina Yañez). *Source* By Authors

**Fig. 5.15 Paediatric
Department—Hospital
21—Queretaro (Mexico)**
(José A. Gonzales). *Source*
By Authors

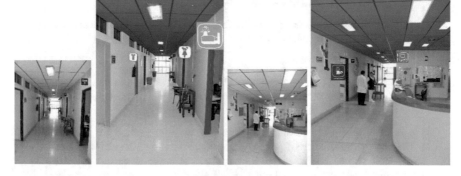

Fig. 5.16 Paediatric Department—Hospital 2—Queretaro (Mexico) Corridors and reception (as they are and the proposals) (Anna Luisa Castillo Bonadies). *Source* By Authors

Fig. 5.17 Paediatric Department—Hospital 2—Queretaro (Mexico)—A reception desk (as they are and the proposals) animated with the story of the monkey Paco, recurring in the Department (Anna Luisa Castillo Bonadies). *Source* By Authors

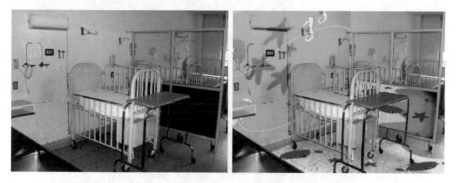

Fig. 5.18 Paediatric Department—Hospital 33—Queretaro (Mexico)—A bed room (as it is and the proposal) (Pilar Román Acosta). *Source* By Authors

Fig. 5.19 Paediatric Department—Hospital 21—Queretaro (Mexico)—A bathroom with an "educational" crab, recommending hygiene and cleanness (Iliana Treviño). *Source* By Authors

Fig. 5.20 Paediatric Department at Hospital of Sondrio (Italy)—The environment of the NMR is due to a contest organized by the company Disney Italia and is based on the characters of the movie *The Little Mermaid* by Disney. *Source* By Authors

Fig. 5.21 Project of a living area in a hospital—The project suggests the use of commercial design furniture of high aesthetical level, and an unusual layout for an hospital (Giancarlo Baffa, Marco Ciffo, Federica Colamaria, Indre Grumbinaite, Monica Izmailovaite, Davide Tettamanzi). *Source* By Authors

which has less look like of a hospital. Paintings stimulated their *Seeking* to know about the paintings and their creator. Their notes were showing their appreciation of the moments of distraction. They'd have less perception of time during their waiting time. A cosy space made them to feel protected with colours, walls, furniture, light and art. They feel more *Care in* this waiting room than others. Later using the Kansei Engineering method was a proof to show patients positive emotion in this waiting room. We talked about it with more details in Chap. 4.

5.4 Waiting Rooms and the Istituto Nazionale dei Tumori Experience

The oncologic centre Istituto Nazionale dei Tumori has a lot of waiting rooms. Almost in every corner of the hospital there is a waiting room. Our focus was more on the main waiting area which we talked about it at Sect. 5.2 and mammography waiting area Sect. 5.3. In both waiting areas cancer patients in different parts of their treatments, pre-cancer, in their treatment process, post cancer to control. All of them were waiting to see a doctor or do an exam. Both waiting areas were designed with consideration of the need of people. Knowing about how people feel in these two waiting areas made us to do some surveys based on KE. The result was people were more satisfied with the mammogram waiting area because of the warm colour, the material of the chairs and the furniture layout. The main waiting area was more successful in furniture layout. The result of both brought us an iterative design process to reconsider people's need and design a new space for them.

5.5 The Life in a Health Care Centre

The life in a Health Care centre can be observed from two points of view: the perceptual environment communicating to the patients and the activities that patients can do.

The first step of empathic process is observing people and learn about their life. We chose a month to be in the hospital to observe cancer patients' life. To learn about how they communicate together and with the environment, how they spend their time during days and nights. Observing them talking to others, looking for a coffee machine during the night, waiting for the doctors to give them some hope, etc. All taught us about their hospital lives. One smile could change their mood from sadness to happiness and viscera a pessimistic word could make them upside down. The observation was recorded and drawing their stories without getting their attention and made us ready for the next step. We learned about the questions we should ask them and the way we need to talk to them in the next step of the design process, knowing people deeper.

Talking to people opened new doors to us to interpret their behaviour thought their words to know their feeling in the hospital, their need and desire. Sometimes we could find some extreme ideas from the patients as well. For example, Alice, 62 years old woman, during her cancer treatment was talking about how happy she is in the hospital, she got everything she need, and she does not expect more than that; Luca, 56 years old man, was angry with whole the experience and unsatisfied with his journey. Talking to people, learning about their character and their experience in the hospital prepared us to calculate data for better ideas at the next step of our design process.

5.5.1   The Environment

The main perceptual characteristics of a typical environment are given by colours, lights, smells and sounds. As discussed in Chap. 2, all of them have a strong influence on the emotional perception by the patients.

5.5.1.1   Colours

Light and color have always been fundamental for the physio-psychological balance of man. Why are so many hospitals still unadorned, badly furnished, badly lit, worse painted? Why are the warm and wise atmospheres present in many of our homes not adopted, why are the intriguing lights used in shops, hotels and offices not used? Finally, why is there almost no trace of the enviable level of quality expressed by Italian design? What model of society do these impersonal, bare, not very functional spaces, generally dominated by colors around white, represent? It almost seems that the idea of the hospital remains as a place to die, where that last suffering was a good opportunity to purify oneself and then that white good represented that stage of purification. Today, fortunately, things are very different, the hospital is no longer the last stop, but a place to heal and where to regenerate to live better. So why are these spaces not designed to intervene, for example, also on our mood, thus guaranteeing the achievement of an easier psychophysical well-being? Surfaces and the environment communicate and their communication heavily affects what the doctor and hospital health personnel will be able to activate. What therapeutic value does a welcoming environment have? It is necessary to intervene in the environments with suitable materials, colors, lights, provisions respectful of the activities and relationships in place. Not a purely aesthetic operation but methods of intervention that are an integral part of a more effective therapy that considers man holistically. So not which color to use but which palette to use. Not single colors but color palettes, custom palettes to build a sort of chromatic identity determined by the place, the structure, the furnishings. A chromatic aggregation that can facilitate the identification of the place. The palette will consist of colors, but also of surfaces, materials, light, signs, advertising, decorative works and everything that contributes to defining the 'landscape' of the place.

From the description of the project *Piano Colore*,

carried on by G. Bertagna and A. Bottolo

for *Fondazione Ospedale Alba-Bra*

(www.fondazioneospedalealbabra.it/progetti/piano-colore/)

The typical colours used in Health Care environments, in particular the public ones managed by national health services, are in general anonymous: pale green, pale blue or grey for the walls, similar colours, or beige for floors and ceilings and so on. Some frames hanging on the walls contain, at best, prints of paintings or posters, and, more often, notices, sanitation recommendations or advertisements about the centre and the services it offers.

Those characteristics constitute an archetype of a Health Care environment, but instead of suggesting healing, recall diseases. An unusual choice of colours, and their use together to some kind of "storytelling" can deeply influence the patients' emotions: in particular, the possibility to "cancel" the persistent atmosphere of the disease can increase *Care* and *Play*, reducing *Fear* and *Panic-Grief*.

In the following proposals are shown to change colours and emotional impact. Obviously, interventions for children's hospitals can add "stories". The figures are self-evident[2] (Figs. 5.14, 5.15, 5.16, 5.17, 5.18, 5.19, 5.20, and 5.21).

5.5.1.2   Lights

The hypothalamus acts on many of the body's functions, for example, regulates emotions, regulates sleep and waking (the circadian rhythm), regulates the intake of food and thirst and also has control of the neurovegetative system. The lighting studied within the nursing environments (where it is not possible to use natural light) is realized so as to go to act and rebalance the natural circadian rhythm of the body which is regulated by natural light thanks to the production of: cortisol for the waking, the body produces it in the presence of light and melatonin that serves to sleep and in the presence of light does not trigger. The biological effects of light can be exploited to improve patient health and well-being.

In many studies showed the importance of light effect in decreasing depression, fatigue, alertness (Ulrich et al., 2004). Using the combination daylight and nightlight suggested to decrease depression and stress. Dunne in 1989 published a paper related to some effect of quality of light in health. He talked about different type of disease causing be fluorescent light. He explains how changing light during the winter might causes stress and depression in people (Dunne, 1989). In Addition, windows in the workplace and having access to daylight are increased satisfaction with the work environment (Boyce et al., 2003; Edwards et al., 2002).

The most common light in the hospital is fluorescent and there is lack of daylights in most of the hospital. Most of the waiting areas have no window and people are sitting between four walls to visit the doctor. Warm colour in radiography waiting area and lightwell in main waiting area are another reason for patients to feel better in these two waiting areas comparing the others.

Non-invasive lighting bodies, with colour temperature similar to that of daylight, can be used in waiting rooms. The market is rich in proposals of great interest: we report some of them (Fig. 5.22), "merging" with the surface of the walls.

The lights can also be used to shape the space, isolating some areas of light, almost to create "bubbles" of personal light: this can be particularly useful in areas where more people stay for a long time, as hospital rooms with more beds. The use of directional and intensity-adjustable spotlights can be suitable for the purpose.

Many therapies (such as radiotherapy or chemotherapy) require the patient periods of a certain duration without being able to move. In such cases, lights with dynamic behaviour can hold in distraction, stimulating the seeking.

Finally, the offers lighting bodies based on LED technologies, characterized by low consumption, high efficiency and a very rich range of colours.

---

[2] The proposals are due to students of Politecnico di Milano, Design Faculty and to ITESM—Campus of Queretaro (Mexico), Faculty of Design. Their names are reported in the figures.

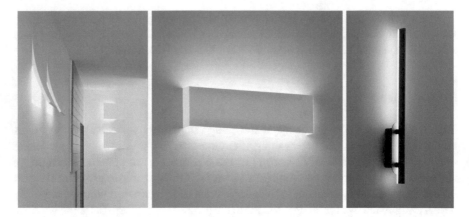

Fig. 5.22 Some wall light—Dodò (Viabizzuno), Righello (Essenzialed), Taranto (KosiLum).
Source We consider the pictures as for "fair use": they have been found on internet, in the sites of
the producers or of the resellers

5.5.1.3 Smells

Lieff and Alper (Lieff et al., 1988) call smell the "most emotional sense", and
Gibbons says, "Our most elusive sense, smell reaches more directly into memory
and emotions than other senses" (Gibbons, 1986). Smell has an important effect
on human behaviour with triggering emotions. Smell can cause anxiety, calmness,
happiness and many other emotions.

Smells bring clear identities to places and connect people emotionally to the
places (Xiao et al., 2020). Human experience of smell has an influential role on the
perception of space.

People in the hospital have a common sense of smell and have effect on people's
perception of the hospital. Smell in the hospital is a combination of drugs, people,
detergents and different chemical materials. Everybody once went to the hospital
has a memory of the smell. Since scents have the quickest flash back to a memory.
Smell "of the hospital" can remind pain, and difficulties and shapes negative feelings
if always stay the same. Smell of cleanness and hygienic place can have a positive
effect on people's perception in the hospital.

Moreover, it is possible to administer scents to get specific collateral effects:
experiments shown that when a vanilla-like fragrance was administered to patients
undergoing MRI scans, they were significantly less anxious (63% less)" reducing
problems in completing the MRI scan (Redd et al., 1994).

But the use of scents in a hospital could influence emotions in a way we could
not control. Let us examine why.

Smell is detected by neurons present into the nasal cavity, and information is
sent to the limbic system, responsible of many emotions. But smell is not directly
connected to some emotions: it is only able to recall past experiences, inducing
related past emotions (Köster, 2002).

Hospital smell recalls illness, painful treatments, stressing diagnoses, and then have a high potential to induce negative emotions.

A scent can negatively affect also medical staff, even when the smell can no longer be consciously perceived.

Those negative emotions slower healing and reduce immunity (Schneiderman et al., 2005).

By contrast, however, certain smells may have a positive effect on well-being, relaxing, increasing alertness, calmness, sociable behaviours and reducing anxiety. Those smells can support recovery from illness, as well as improve the performances of the staff (Schneiderman et al., 2005).

But the fact that smells recall emotions related to past experiences, and so the same smell positive for somebody could be negative for other makes unpredictable the impact on a single person.

5.5.1.4   Sounds

Sounds in the hospital are usually unwanted, the sound of conversations between doctors, nurses, patients and their friends and family, sound of carrying things around and sometimes structures. Hearing other patients moaning and suffering impacts on other patients' emotions. Sound of hospital considers as a negative term. Music as a useful distraction is recognized to effect positively on patients and nurses.

Iyendo believes "music as a complementary medicine for improving health care". (Iyendo, 2016). His study showed that soothing music reduces stress, blood pressure and post-operative trauma and improves the hospital experience. Ulrich (Ulrich, 2004) emphasizes at useful distraction theory in the hospital which advocates music, bird and water sound help healing. Department of Health highlighted the positive effect of music in treating depression as well.

Possibly, music can be used in a targeted way to stimulate specific emotions. In Chap. 6, we present recent researches and experiments able to correlate the arousal of specific emotions with a specific listening.

5.5.2   Activities

In a Health Care environment, patients are torn from their usual life and daily affairs, and time passes slowly; for this reason it is very useful to propose activities to keep mind and body engaged.

Besides the usual areas for visits and for watching TV other activities can be organized. We report in the following some of them, carried on at Istituto dei Tumori in Milan.

5.5.2.1   Book Crossing

The waiting areas related to specific exams and therapies (the ones devoted to hospitalized patients) have been equipped with shelves, on which everybody was free to leave or to get a book.

The initiative, simple and cheap, has been always successful.

5.5.2.2   Writing a Diary

Various projects have been carried on, for adults and children. The proposal for adults has been used simply to write every day some notes, keeping a track of the thoughts and of their evolution. More effort has been carried on for children.

5.5.3   My Backpack—*A Companion for Children*[3]

The moment of hospitalization is an extremely traumatic event for children: the relationships that they established with the outside world change, the trust in parents ability to protect them decreases, and instability arises.

Such an instability can be mitigated through adequate communication, which does not deny the negative state of health, but creates positive awareness, capable of explaining the reasons for hospitalization, for medicines, and allowing to verify the course. The project is specifically oriented to the preschool age. The idea is that children can build a personalized fairy tale, telling about themselves in the form of a game.

The underlying principles are:

- the perception of the disease as injustice, with an external "culprit";
- a sort of narration, operated by children, with the antagonist (the disease), the hero (the doctor), the helpers (friends or nurses), the magical means (medicine).

The game tools are contained inside a fabric backpack, full of pockets and cases, which creates a surprise effect, with the discovery of the different games. The white canvas backpack can be customized with felt-tip pen drawings. Inside the backpack, there is a "path" (referred to as "my day") that allows children to build the progress of the day from waking up to sleep, with the help of cards/icons of the various possible activities. In this way, the child gets awareness of the hospital experience. It is also a way to plan the day. Icons allow to characterize the dominant emotional state of the day. Gaming activities focus on *colouring* (for the figures that appear in the story), *interact* (to become familiar with the various medical equipment, with drugs, and become aware of the illness), *create* (develop stories, to be shared with friends and parents) (Fig. 5.23).

---

[3] The project is due to G. Cistino and V. Prosperi.

Fig. 5.23 *My Backpack*—The backpack with pockets, booklets and pencils; the "my day" area; the personalized fabric, the cards of the various characters, the icons of equipment and drugs. *Source* By Authors

Fig. 5.24 *My Doctor-suitcase*—A project specifically designed against *Fear*, for children 6–10 years old. A cardboard suitcase containing many objects, a booklet and provided with a secret compartment. Two characters, Milla and Tom, "interact" with the child, in different sections (Introducing Milla and Tom, My Identity Card, What happened today), (Draw your room, Draw your illness and so on) (project by S. Furlani and C. Taddia). *Source* By Authors

Fig. 5.25 *Fantanimando*—Based on the idea of the game *Più e Meno* (Munari et al., 1970), for children 8–12 years old. Acetate sheets with images of hospital elements (environments, furniture, people, equipment, etc.) can be superimposed to build scenarios. The game can be customized through "comic clouds" and acetate sheets on which it is possible to draw with felt-tip pens. (project by M. Casarotto and G. Maffeis). *Source* By Authors

Other similar projects have been carried on, we refer simply in the following figures: details on the projects can be found in (Maiocchi, 2010) (Figs. 5.24 and 5.25)

5.5.4 *Gardens of Wellness*

Many plants are the basis of active ingredients used in medicine. The construction of "wellness gardens", that systematically collect these plants and show them to patients, is common practice in many hospitals. One of these wellness gardens has been built on the rooftop terrace of the INT in Milan. The project, carried out by the Department of Agricultural Engineering of the Politecnico di Milano, consists of a certain set of areas in which the plants are grouped according to their characteristics and related curative aspects, also with the aim of being able to organize didactic paths (Senes et al., 2012).

Many authors support the idea that wellness gardens can help to cure, contributing in different ways;

Fig. 5.26 *Wellbeing Gardens*—The gardens on the rooftop terrace of the Istituto Nazionale dei Tumori in Milan. *Source* Pictures are a re-elaboration of images publicly presented in some conference, and a reference to those presentation is present in the text

- distraction and involuntary attention (Attention Restoration Theory, Kaplan, 1989);
- reased stress, acquisition of a "sense of control", positive distractions (Stress Centrality Theory, Ulrich, 2006);
- increase immune efficiency, promote a favourable hormonal profile to cope with stress, reduce cortisol and adrenaline production (Positive Thinking Theory, Huppert et al., 2011);
- landscape, flowers, sight are positive "qualifiers" that direct our response and our behaviour; they activate optimism, hope and the desire to react (Damasio, 1994) (Fig. 5.26).

Fig 5.27 Shows the organization of the Wellbeing Gardens at INT

5.5.4.1 Cooking for Health

Near to the main building, an old refurbished farmhouse, *Cascina Rosa* (ruled by prof. Berrino), houses some structures of the Istituto dei Tumori in Milan (Fig. 5.28).

The *Cascina Rosa Campus* hosts research in the field of epidemiology, prevention, public health and organizes numerous training initiatives open to the public: the covered topics range to primary prevention to behaviours reducing the risk of onset of a tumour. The natural cooking courses organized by the associations operating in the Cascina involve hundreds of people every year. The products used to come directly from the synergistic garden created specifically in the Cascina (Fig. 5.29).

5.6 Final Remarks

Many other courses and events are organized, such as musical concerts, courses on painting and drawing and so on.

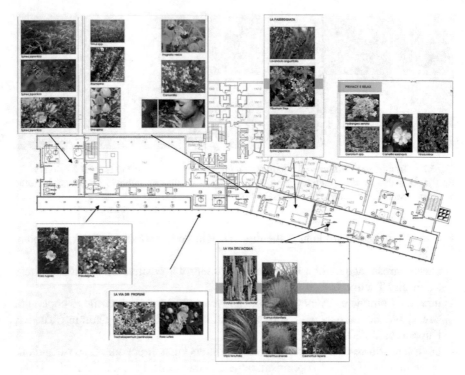

Fig. 5.27 *Wellbeing Gardens* **at Istituto Nazionale dei Tumori in Milan**—The plan. *Source* Pictures are a re-elaboration of images publicly presented in some conference, and a reference to those presentation is present in the text

Fig. 5.28 Cascina *Rosa* (Istituto Nazionale dei Tumori in Milan). *Source* Pictures have been found in the websites of the referred organization, and we consider them as for "fair use"

■ **Corso di base**

È possibile partecipare a questo corso di 3 lezioni ogni mese, il giovedì - tre giovedì consecutivi - oppure il sabato (mattina e pomeriggio) e la domenica (mattina). Imparerete l'arte della prevenzione e della cura del cibo, le proprietà terapeutiche degli alimenti e il loro utilizzo appropriato che tenga conto del clima e della stagione, della costituzione della persona e della condizione fisica in cui si trova.

Argomenti trattati dalle nutrizioniste e dai cuochi:

• Le regole della buona cucina
• L'equilibrio nella composizione del pasto
• Preparazione dei cereali integrali
• Piatti proteici con pesce o legumi
• Zuppe e minestre squisite e pratiche
• Cotte o crude? Tante idee sulle verdure
• Come fare un dolce sia buono che salutare
• Evitare gli errori nutrizionali più comuni
• Scelta degli ingredienti, dei materiali
• Lettura delle etichette

Orario:
giovedì dalle 17.30 alle 20.30
Oppure nel fine settimana:
sabato dalle 10.30 alle 13.00 e dalle 15.30 alle 19.30
domenica dalle 10.30 alle 13.00

■ **Per prenotazioni ed iscrizioni:**
Salute Donna Onlus:
NUMERO VERDE 800 223295
Mercoledì dalle ore 16.00 alle 18.00
Tutti i giorni è possibile lasciare un messaggio
ed un recapito telefonico per essere richiamati.
Oppure via e-mail a:
corsicucina@salutedonnaonlus.it

I Corsi sono aperti a tutti

■ **Potete effettuare il pagamento:**
• Tramite bonifico bancario:
BANCA PROSSIMA Filiale 5000
Piazza Paolo Ferrari, 10 - 20121 Milano
Cod. IBAN
IT11V0335901600100000002935
intestato a: Associazione Salute Donna onlus
Via Venezian, 1 - 20133 Milano
• oppure Conto Corrente Postale: **119206**
intestato a: Associazione Salute Donna onlus
Via Venezian, 1 - 20133 Milano
IMPORTANTE: indicare sulla causale
del versamento **"Corsi Cucina"**

Fig. 5.29 Natural *cooking courses at Cascina Rosa. Source* Pictures have been found in the websites of the referred organization, and we consider them as for "fair use"

References

Boyce, P., Hunter, C., & Howlett, O. (2003). *The benefits of daylight through windows*. Rensselaer Polytechnic Institute.

Damasio, A. (1994). *Descartes' error: Emotion, reason, and the human brain*. Penguin.

Dunne, A. (1989). *Some effects of the quality of light on health*. J. Orthomol. Med 4.

Edwards, L., & Torcellini, P. (2002). *A literature review of the effects of natural light on building occupants* (Technical report). National Renewable Energy Laboratory.

Gibbons, B. (1986). The intimate sense of smell. *National Geography Magazine, 170.*

Huppert, F. A., & Linley, P. A. (Eds.). (2011). *Happiness and well-being*, Routledge.

Iyendo, T. O. (2016). Exploring the effect of sound and music on health in hospital settings: A narrative review. *International Journal of Nursing Studies, 63.*

Kaplan, R., & Kaplan, S. (1989). *The experience of nature: A psychological perspective*. Cambridge University Press.

Köster, E. P. (2002). The specific characteristics of the sense of smell. Iin *Olfaction, Taste, and Cognition*. Cambridge University Press.

Lieff, B., & Alper, J. (1988). Aroma driven: On sense the trail of our most emotional. *Health,*
    20(12), 62–65.
Maiocchi, M. (Ed). (2007). *La comunicazione emozionale nell'ambiente ospedaliero*, Maggioli.
Maiocchi M. (Ed). (2008). *Design e Comunicazione nella Sanità*, Maggioli.
Maiocchi, M. (Ed). (2010). *Design e Medicina*, Maggioli.
Munari, B., & Belgrano, G. (1970). *Più e meno*. Corraini.
Redd, W. H., Manne, S. L., Peters, B., Jacobsen, P. B., & Schmidt, H. (1994, July/August). Fragrance
    administration to reduce anxiety during MR imaging. *Journal of Magnetic Resonance Imaging.*
Schneiderman, N., Ironson, G., & Siegel, S. D. (2005). Stress and health: Psychological, behavioral,
    and biological determinants. *Annual review of clinical psychology, 1.*
Senes, G., & Fermani, E. (2012). Healing gardens: The garden on the terrace of the National
    Oncology Institute in Milan (Italy). In *Global Congress for Qualitative Health Research* (p. 29).
    Vita e Pensiaro.
Ulrich, R. S., Zimring, C., Joseph, A., Quan, X., & Choudhary, R. (2004). The role of the physical
    environment in the hospital of the 21st century: A once-in-a-lifetime opportunity. *The Center for*
    Health Design.
Ulrich, R. S. (2006). Essay: Evidence-based health-care architecture. *The Lancet, 368,* 38–39.
Xiao, J., Tait, M., & Kang, J. (2020). Understanding smellscapes: Sense-making of smell-triggered
    emotions in place. *Emotion, Space and Society, 37.*

Chapter 6
Beyond Design

6.1 A Preliminary Note

Beside the actions we have experienced working on architectural spaces, on furniture and procedures, we carried on some researches on the possible use of music and of videos to positively influence patients' emotions.

We did not yet apply the results to health care environments, but we refer here about the principles of the studies and the obtained results and it has a potential to use in health care centres.

Moreover, we strongly believe that smells should be investigated too.

6.2 Experiences with Music

Sounds are an essential component of an environment: the experience of whoever enters an anechoic chamber is that of a total estrangement, due to the complete absence of reference noises.

In a typical health care environment, the only noises you can catch are the ones related to people, staff, movements and acoustic signals. All those sounds are mixed and reverberated by generally large, high and bare environments. What you hear is mainly related to the subdued and indistinct shouting in the waiting rooms, to the shuffling of the nurses' slippers and the rolling noise of trolleys and beds being moved in the corridors, to the acoustic warnings that inform people waiting that their turn has arrived.

All those sounds are typical of the health care environment and recall just illness.

We could think to introduce background music in all places where speaking and listening activities are not relevant (i.e. not for a visit, or in the parlours or in the hospital rooms): waiting rooms, passageways, areas where patients are forced to spend time during a therapy (think for example, radiotherapy, or other cases in which there are treatments of a certain duration, in the absence of persons).

6.2.1 The Study

In this paragraph, we report the results of a research carried on in cooperation between the Politecnico di Milano and the Conservatorio G. Verdi di Milano, aiming to explore and define the technical characteristics of the music and their capability of rising emotions (Maiocchi & Rapattoni, 2017).

We used as a reference the Panksepp's model for the emotions and explored the reactions of many persons through a Kansei method, going through the following steps:

- we selected twenty-four pieces from those supposed rich of an emotional connotation and proposed to listen them (recorded on a movie, with black video and identified just by a number, without title and author) to about fifty persons, asking them to associate each piece to emotional adjectives; the following pieces were selected:

1. F. Liszt—Liebestraum n. 3	15. S. Gubaidulina—Introitus
2. H. Berlioz—Symphonie Fantastique Mov. 5	16. S. Rachmanonoff—Little Red Riding Hood
3. A. Vivaldi—from "Il Farnece"	17. I. Stravinsky—The Rite of the Spring
4. D. Shostakovich—Polka from "The Golden Age"	18. G. Cardini—Concerto for Demetrio Stratos
5. R. Wagner—Ride of the Valkyries	19. J. Brahms—Trio in B sharp Op. 8
6. I. Stravinsky—from "The Firebird"	20. W. A. Mozart - Requiem—Confutatis Maledictis
7. A. Pärt-Silouans Song	21. G. Mahler—Adagietto from Symphony n. 5
8. G. Gaslini—Soundtrack of "Deep Red"	22. C. Debussy—Prelude n.1 from Minstrels Vol. I
9. J. Desprez—La Déploration sur la Mort d'Ockeghem	23. R. Wagner—Sigfried Funeral March
10. L. Bernstein—Candide (Overture)	24. W. Lutoslawsky—String Quartet—Main
11. H. Cowell—Dynamic Motion	
12. P. Glass—Einstein on the Beach	
13. D. Shostakovich—Piano Concerto n. 2 (2nd Mov)	
14. V. Tormis—Curse Upon Iron	

- over nine hundred different adjectives were collected; each of the adjectives has been ascribed to a class corresponding to the selected emotional systems (for example, *romantic* in Care, *distressing* in Fear, *depressed* in Grief, *playful* in Play, *impetuous* in Rage, *incomprehensible* in Seeking), discarding the ambiguous adjectives;
- organizing the data into a table, we observed that many pieces were definitely associated to a specific emotional system, while other presented some ambiguities; we discarded the last ones, reducing the pieces to eighteen;
- then a musical analysis examined the technical characteristics of each piece, in order to associate them to the emotional systems (for example, irregular rhythms, fast time, microtones, dissonances and other are associated to *Seeking*; major tonality, tonal harmony, *legato*, uniform texture and so on, to *Care*): all the results are discussed in detail in (Maiocchi & Rapattoni, 2017).

Table 6.1 Association between emotional systems and some pieces of music

Emotion	Music
Care	1. F. Liszt—Liebestraum n. 3
Fear	V. Tormis—Curse Upon Iron
Rage	R. Wagner—Ride of the Valkyries
Panic/Grief	A. Vivaldi—from "Il Farnace", *Gelido in ogni vena*
Play	D. Shostakovich—Polka from "The Golden Age"
Seeking	W. Lutoslawsky—String Quartet—Main

By the way, we did not find any piece of music able to raise *Lust*.

Following the above results, we are able to select the proper kind of music as a background, to raise *Care* and *Play* to cheer up patients, or *Seeking* to positively distract them.

After the study, we selected six pieces typical for each of the six selected emotional systems and verified the emotional response to a wider sample of listener, always confirming the results.

For who could be interested to directly verify the association between emotions and music, we propose the listening of those final six, easy to be found on internet, reported in the following list, together with the associated emotional system (Table 6.1).

Moreover, through the analysis of many other pieces, we were able to describe the musical parameters able to characterize music from the point of view of the emotions' arousal.

Tables 6.2 and 6.3 (following pages) reflect the results of the research (Maiocchi & Rapattoni, 2017). The parameters taken into account are either *conventional* (i.e. the ones we can find indicated on the musical scores) or *unconventional*, such as *density* or *complexhy* (Sciarrino, 1998). Each parameter is declined in its possible forms and associated to the raised emotions. Such an association has been done through Kansei engineering evaluations. Moreover, the forms of the parameters have been associated also to the perceptual phenomena as described by Ramachandran (P = Peak Shift, G = Perceptual Grouping and Binding, C = Contrast, I = Isolation, S = Perceptual Problem Solving, A = Abhorrence of Coincidences/Simple Interpretation, Sy = Symmetries, R = Rhythm and Repetition, B = Balance).

As we already said, we have not carried on experiments on music in the hospitals. But we consider it is a further chance for orienting positive emotions.

Experiments could be carried on using a background music in the waiting rooms, able to increase *Seeking* (distracting patients from negative thoughts), *Care* and *Play*.

For example, if we desire the arousal of a *Seeking* mood, we can observe that the elements characterizing such an emotion are musically correlated to the following elements:

Table 6.2 Association between emotional systems and musical conventional parameters

Conventional parameters			Seeking	Rage	Fear	Care	Grief	Play	Ramach
Rhythm	Regular			●	●	●	●	●	●	G
	Irregular		●						CAG
Tempo	Slow					●	●		GR
	Fast		●	●	●			●	
	Accelerating			●	●			●	C
	Changing		●	●				●	C
	Distended					●	●		B
Melody	Continuous				●	●	●		G
	Jumps		●	●				●	C
	Wide intervals								S
Harmony	Tonal	Regular Major		●		●		●	G
		Minor			●		●		G
		Irregular Changing	●						C
		Microtones	●		●				S
		Dissonant	●	●	●				C
		Clusters		●	●				P
	Atonal	Microtones	●		●				S
		Dissonant	●	●	●				C
		Clusters		●	●				P
	Chromatisms		●						G
Timber	Uniform	High pitch				●		●	I
		Low pitch	●	●	●	●	●	●	I
		Dull sounds	●	●	●		●		I
		Brilliant sounds	●			●		●	I
	Complex	Ascending pitches							C
		Descending pitches		●			●		C
		Mixed	●		●			●	C
		Dull sounds		●	●				C
		Brilliant sounds					●		C
		Mixed	●	●	●			●	C
Dynamics	Constant	Piano				●	●	●	I

(continued)

Table 6.2 (continued)

Conventional parameters			Seeking	Rage	Fear	Care	Grief	Play	Ramach
		Forte		●		●	●	●	I
	Floating	Crescendo	●						C
		Diminuendo					●		C
		Changing	●	●	●			●	C
Touch	Staccato			●	●	●		●	I
	Legato					●	●		G
	Re-beated				●				P
Texture	Regular	Uniform	●		●	●			B
		Variations						●	G
		Repetitions	●		●				R
	Irregular	Instable	●						S
		Open geometries	●						S
		Repetitions			●				R

Table 6.3 Association between emotional systems and musical unconventional parameters

Conventional parameters			Seeking	Rage	Fear	Care	Grief	Play	Ramach
Density	Constant					●	●	●	B
	Variable	Accumulation	●	●	●				C
		Rarefaction					●		C
		Mixed						●	C
Complexity	High			●		●			S
	Low			●		●	●	●	B
	Changing	Stretching	●	●	●				S
		Spreading				●	●		
		Mix						●	C

- *constant rhythm*, sometimes alternating with break of the previous structures;
- *slow tempo*, especially in the face of well-defined musical themes, often with sudden passages at fast tempos, strongly contrasting the created state;
- *melody* seems *not relevant*, but a clear and well-defined melody does not help;
- the same for *harmony*;
- *high variable timbres*, with alternations of dark and low tones to bright and high tones;
- *variable dynamics* and alternation of sound intensities;
- *variable pitches* and superimposed musical registers are present, and in particular relevant when, with a single playing instrument, there is no possibility of significant timbral variations;

- presence of *repetitions*, sometimes irregular;
- open and often *unstable structure*;
- *variable density* in the accumulation and rarefaction of musical themes;
- *high complexity*, with strong irregularities, even overlapping.

Obviously, these elements are neither necessary nor sufficient; simply, their relevance in listening has been observed, and it is therefore assumed that they contribute to making Seeking perceive.

Examples of musical pieces corresponding to those parameters are:

- Igor Stravinsky—*The Firebird*. The piece constructs a presentation of low tones, with regular and slow rhythms; clarinets and double basses muted a dark theme: a waiting situation, as if the listener was looking around waiting to understand what will happen. Then the sudden appearance of the fast triplets of the strings and the woods, with their bright tones, herald the arrival of the firebird, to satisfy the wait, as if the solution had been reached; immediately afterwards the search begins again, until you reach a more defined, more evident bird entrance, which is completed with the bird's dance. Seeking is recreated in the listener from the dark background, and satisfied by the entry of brilliant timbres.
- Philip Glass—*Einstein On the Beach*. Here the mechanism that creates the Seeking impression is completely different. The listener does not immerse himself in a narrated Seeking situation, he does not interpret the narrated Seeking, but tries to interpret the musical structures that are proposed, and in this he finds the musical environment in which to operate his own path of personal research, by Seeking to understand what the intent author proposes.
- Josquin des Prez—*Qui habitat*. Seeking's impression is probably linked to the many voices that follow each other. In this case, we should think that this emotion is strongly present in many typical pieces of counterpoint.
- Giancarlo Cardini—*Concert for Demetrio Stratos—Novelletta* (1979). The compositional mechanism finds expression in the interpretation of spots on the score. Knowledge of the fact could influence Seeking's perception, which is therefore inevitably influenced by processes in the cerebral cortex, which searches for structures.
- Sergej Rachmaninov—*Study Op. 39 n. 6*. Seeking's perception is given by the uncertain character of the beginning, characterized by a variety of rhythmic situations, emotional mobility and changes in dynamics.
- Franz Liszt—*Mephisto Waltz No. 1*. In particular, at the beginning, there is a tension of research and exploration of different musical patterns.

We provide in the following other examples for *Care* and *Play*.

For *Care*, the musical parameters are characterized by regular and constant rhythm; slow or andante tempo; continuous and regular melody; major tone harmony; uniform timber, not gloomy; constant dynamics; very close sonority; regular uniform texture; constant density; low complexity.

Examples of musical pieces are:

- Franz Liszt—*Liebestraum n. 3—Nocturne and Liebeslied*. The melodic width, the tonality, the wide slurs and the constant figuration of the accompaniment make the listener feel softly enveloped by the music, which certainly suggests suffering from *Care*.
- Dmitrij Shostakovich—*Piano Concerto No. 2: II. Andante*. Written for the academic diploma of instrumentalist of his son Maxim, it seems a wish of joy to the son; the piano, accompanied only by strings and a horn, results in a sweet song, sometimes melancholy, with delicate sonority.
- Franz Schubert—*Piano quintet in a major, d.667 "The trout"—iv. theme (andantino) with variations 1–6*. The tonality, the modulations and the timing are linked to *Care*, but there are very present, perhaps even more, aspects of *Play* (which we will see later), due to the punctual rhythm and the presence of trills.
- Sergej. Rachmaninov—*Rhapsody on a Theme of Paganini, Op. 43*. In spite of the fact that there are melodic and performance elements that recall *Care*, one perceives excesses of legacy and breadth that rather suggest a celebratory will, told rather than felt, with a hint of falsity.

Finally, for *Play*, the musical parameters are: brilliant rhythmic patterns, sometimes askew, changing; generally fast tempo, with accelerations; melody often jerky and irregular; major mode harmony, dissonances; brilliant timbres, percussive timbres; often floating dynamics; sonority increasing towards the top, with glissando, staccato and repeated notes effects; "ungrammatical" textures, variations; variable density; changing complexity.

Examples of musical pieces are:

- Dmitrij Shostakovich—*Polka, from The Golden Age, op. 22b*. The theme is presented with dotted and detached rhythms reminiscent of childhood play; even the timbres have a touch of playfulness, the joke is implicit in the form and the desired "ungrammaticals" further enhance a childish naivety.
- Wolfgang A. Mozart—*12 variations for piano on Ah, vous dirais-je, Maman*. Already the idea of variation has something of a game; in particular, Mozart's version is now full of contrapuntal games, now of *acciaccature* support; the graceful display often adds a colour of *Care*.
- Francesco B. Pratella—*The War*. The choice of the major key would also be congruent with the aggressiveness suggested by the theme, which however is treated in a playful and light-hearted way, as if war were a game, and in this the interpretation of the theme is completely coherent, not the spirit literary future is of the time.
- Leonard Bernstein—*Candide—Overture*. There are many elements that lead back to the game: the presence of scales, the tonal and harmonic contrasts, and the presence of the xylophone that increases the playful image.
- Johann S. Bach—*Prelude and Fugue No. 4 in C minor, BWV 849, from the well-Tempered Harpsichord*. The interpretation of the game, present after a few minutes from the beginning, would seem linked to the counter-punctual game of escape, and therefore perhaps typical of all escapes; certainly, the performance heard, by Glenn Gould, enhances the characteristics of *Play*.

- Josquin des Prez—*Qui habitat*. Despite being a canon of "only" 24 voices, the density of the sound encourages multiple resonances: a brilliant sound with no precise direction, but in constant motion (see also *Seeking*).
- Claude Debussy—*I vol. preludes n. 12 "Minstrels"*. The detachments and the crooked rhythms, with particular accentuations and numerous repeats, sudden crescendo, attribute a continuous play movement to the piece, to which the significant presence of dissonances contributes. See also Prelude 6 II volume.
- Aaron Copland—*The Cat and The Mouse*. The speed changes accentuate the descriptivism of the bra-no, which results in a representation of the game rather than the playfulness of the music.
- Sergej Rachmaninov—*Etude—Tableau Op 39 No. 6 Little Red Riding Hood*. Also in this case the elements of *Play* are linked to the rhythms, fast and accelerated, with a certain descriptive will of the subject.

6.3   Experiences with Movies

While music is not a carrier of semantics in itself (Maiocchi & Rapattoni, 2017), and a meaning can be found just in the lyrics of the songs (or in the story told in an Opera), a movie is a more complex artefact, joining images, words, music and a story. All of them can contribute to raise emotions, and their combination can be really powerful. Researches have been carried on (Radeta et al., 2014), by studying the emotional reactions of persons watching movies, using Kansei engineering methods.

The research has shown that:

- It is possible to describe the timeline of the "movements" of each emotion along the development of the movie (an example is shown in Fig. 6.1), and the emotional reactions are the same for the same scenes for different viewers;

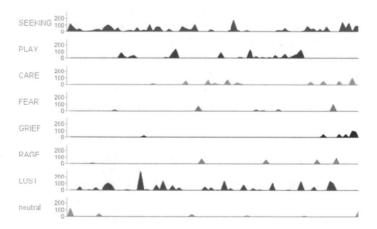

Fig. 6.1 Mean value and authors' accordance on felt emotion per scene. Movie: *Nine 1/2 Weeks* by Adrian Lyne, with Mickey Rourke, Kim Basinger (1986). *Source* By Authors

- each movie can be characterized by few main emotions, and the same emotion can play different roles, according to the presence of others; for example, the dominance of *Seeking* in *Birds* movie (see Fig. 6.2) is biased by the presence of *Fear*, resulting so in a negative emotion, while the same dominance in *Pretty Woman* movie, joined to *Care* and *Play* can lead to positive emotions;
- *Seeking* seems essential to the movies, and each of them shows a high presence of such an emotion (see Fig. 6.3).

The perceptual elements leveraging emotions are many and different; f. i.:

- *Care* in *Singin' in the rain* movie can be related to music and to the story;
- *Care* in *Bambi* movie is related to the story and the drawings;
- *Rage* in *The Incident* movie is due to the direction and the excellent interpretation by Tony Musante;
- *Fear* in *Psycho* movie is for sure related to the story and to the direction;
- *Seeking* in *Birds* movie is influenced by the absence of a musical soundtrack and so on.

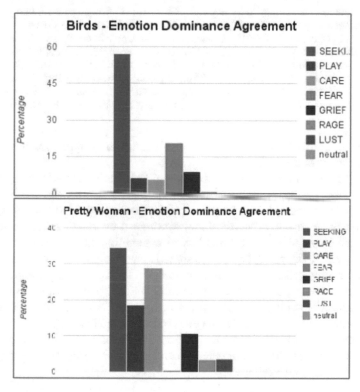

Fig. 6.2 Dominant emotions in *Birds* by Alfred Hitchcock with Rod Taylor, Tippi Hedren, Suzanne Pleshette (1963), and in *Pretty Woman*, by Garry Marshall, with Richard Gere, Julia Roberts (1990). *Source* By Authors

(a)

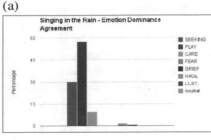
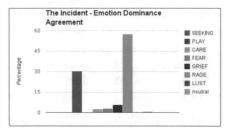
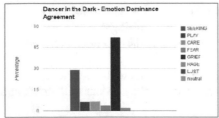
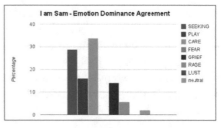

Fig. 6.3 Dominant emotions in some movie. (**a**) First part; (**b**) Second part; (**c**) Last part (From left to right, up to down: *Singin' in the Rain* by S. Donen, w/ G. Kelly, D. O'Connor [1952]—*The Incident* by L. Peerce, w/ T. Musante, M. Sheen [1967]—*Dancer in the Dark* by L. von Trier. w/ Björk, C. Deneuve [2000]—*I am Sam* by J. Nelson. w/ S. Penn, M. Pfeiffer [2001]—*Bambi* by J. Algar [1942]—*Night of the Living Dead* by G. Romero. w/ D. Jones, J. O'Dea [1968]—*Bitter Moon* by R. Polanski. w/ H. Grant, K. Scott Thomas [1992]—*Psycho* by Alfred Hitchcock. w/ A. Perkins, J. Leigh, V. Miles [1960]—*Saw* by J. Wan. w/ C. Elwes, L. Whannell [2004]—*Schindler's List* by S. Spielberg. w/ L. Neeson, R. Fiennes, B. Kingsley [1993]). *Source* By Authors

It is quite difficult to use a movie in a hospital, due to the typical length in time,[1] but it is possible to use some short videoclips.

An interesting proposal is due to M. Aureggi (2010): the idea is to create "video-mosaics", thanks to the possibility of finding, through the Internet and YouTube, huge quantities of video materials to be reused; the collection of video "splinters" could in fact constitute an interesting database, catalogued by emotional characteristics, and by topics covered, with the possibility of assembling them, even automatically, an building small videoclips oriented to specific environments of use. The idea was tested by creating well-defined fixed structures of a short cinematographic narrative, with a well-defined development and with rules for the selection of the "splinters".

Trial carried on with the support of the Istituto Nazionale dei Tumori of Milan produced some videoclips, such as the ones indicated in Table 6.4.

As an example, we report here the storyboard of the first video-mosaic produced, *Inside a Labyrinth of Mirrors*,[2] oriented to teenagers discovering to have cancer; the narrative structure is the following:

---

[1] Nevertheless, we already mentioned the use of cartoons during ultrasound exams to children at Pausilipon Children Cancer Hospital in Naples.

[2] Designed and produced by M. Mandelli, E. Medolago and M. Milani.

(b)

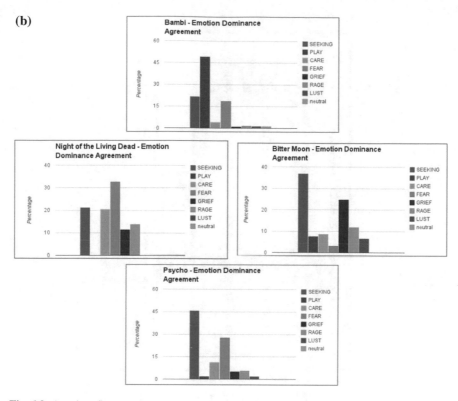

Fig. 6.3 (continued)

The various phases have been joined to different soundtracks, free from copyright, got from internet. Music was chosen according to the represented emotions (Figs. 6.4, 6.5, 6.6, 6.7, 6.8 and 6.9).

After the creation of the video-mosaic, the staff who assist adolescents analyzed the reactions in viewing the video, involving, with the following remarks:

- the video reflects the personality of adolescents, has the right language (the metaphorical and evocative splinters, the use of cartoon graphic style or 3D animation);
- the phases of the video (the five sub-themes) are recognizable and correspond correctly to phases of the disease (even recurring in a non-linear way over time);
- the overall tone of the video is effective, sometimes pressing; the video stimulates reflections and tells the reality in an intense way;
- the positive tone of the ending is seen as an incitement, healing means integrating the disease (even if we are not always able to defeat it);
- the video appears particularly suitable also for those who have already faced the disease; the vision can be a pretext to re-elaborate the experience and make the children aware of the intense personal growth they have faced.

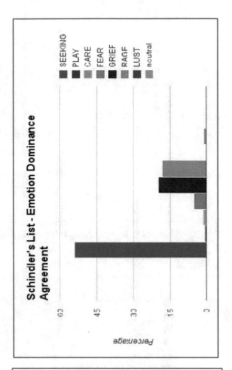

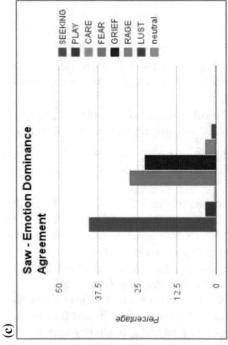

Fig. 6.3 (continued)

Table 6.4 Videoclips produced with automatic procedures

Title	Subject	Emotion
Inside a Labyrinth of Mirrors	Identity	Care
Between Fairy Tale and Reality	Cancer as an enemy to be defeated	Rage against cancer
Beautiful	Body mutation	Care
Help	Overcoming fear	Against Fear
Free from Fear	Overcoming fear	Against Fear
A Lifestyle for Cancer Prevention	Nutrition	Educational—Care
You can Prevent It	Smoking	Educational—Care
Beyond the Obstacle	After Diagnosis	Against Fear
Smoked Omelette	Smoking	Educational—Care

Fig. 6.4 An introduction—Teenagers—Soundtrack: Arpamb (Kreyk)—Illambient, (Kreyk). *Source* By Authors

Fig. 6.5 Discovery of the disease—Soundtrack: Experiment (J.BlascoTX). *Source* By Authors

Fig. 6.6   Accept the physical changes—Soundtrack: Beauty (Tryad). *Source* By Authors

Fig. 6.7   Cope with fear—Soundtrack: The exit, please? (P.Ref). *Source* By Authors

Fig. 6.8   Search and accept help—Soundtrack: Breeth (Tryad). *Source* By Authors

Fig. 6.9   Victory—Soundtrack: The Sailor (72Zh)

References

Aureggi, M. (2010). Videomosaici. In M. Maiocchi (Ed.), *Design e Medicina*. Maggioli.
Maiocchi, M., & Rapattoni, M. (2017). *La musica del sentire. Le relazioni tra suono ed emozioni.* Luca Sossella editore.

Radeta, M., Shafieyoun, Z., & Maiocchi, M. (2014). Affective timelines towards the primary-process emotions of movie watchers: Measurements based on self-annotation and affective neuro-science. In J. Salamanca, P. Desmet, A. Burbano, G. Ludden, & J. Maya (Eds.), *9th International Conference on Design and Emotion* (pp. 679–688).

Sciarrino, S. (1998). *Le figure della musica da Beethoven ad oggi.* Ricordi.

Chapter 7
Driving a Health Care Environment Evolution Through Emotional Design: A General Model

7.1 Goals and Boundaries

7.1.1 Note

In this chapter, we will examine some principles and practices, typical of Total Quality approach, to introduce changes in a health care environment, with the needed consciousness on the following questions:

- Where, how and why to introduce changes in the centre?
- What is the resulting effect of the changes?

Those questions imply to have a clear general model of what an improvement produces and of how to measure it.

A health care organization, as any other organization, operates within a context changing continuously:

- the medical knowledge evolves, providing new models, therapies and drugs;
- the technologies change, providing new methods and tools for exams and therapies;
- the lifestyle evolves, making some illnesses disappear and others spread;
- the social organization evolves, dragging changes in the public health support;
- the economy changes, deeply influencing the costs and the distribution capability of health care support and so on.

So, also the health care organization must evolve, with two fundamental goals, to improve the performances:

- according to the above changes: therapies evolve, as well as social prevention (f.i. early exams, vaccines, new drugs, etc.); public health care coverage spreads, and then the number of potential patients grows;

Fig. 7.1 The feedback
model. *Source* By Authors

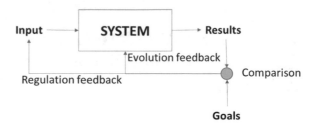

- regardless of the changes in the context given a stable knowledge and therapeutic protocols, it is possible to make the environment evolve to increase efficiency and effectiveness, including a cost reduction.

A general model helping in a proper control of the evolution of an organization refers to the feedback concept: we model an organization as a black-box, with inputs and outputs. We define some kind of measure of the outputs and define quantitative goals for them; a comparison between outputs and goals allows to change some previous condition, in order that results better match the goal.

A typical example is constituted by a heating system: Given a flow of fuel that feeds the system, the heated environment reaches a certain temperature; suppose our goal is to reach the temperature of 20 °C: if the measured temperature is lower, a feedback action will increase the fuel flow, and if it is higher, it will decrease. In this way, the environment will stabilize on the desired temperature.

This feedback model allows to reach a stable situation, but our goal is different. We need to go on for a continuous improvement of the performances, so, we want to continuously improve the level of the goals.

The application of such a kind of model to an organization suggests to better specify it, as in Fig. 7.1.

Beside the *regulation feedback*, which aims to reach the goal in a stable situation, we introduce an *evolution feedback*, changing the "internal structure" of the organization, allowing us to tend to more challenging goals.

So, while a doctor, who is monitoring the status of a patient measuring some medical parameters, changes the dosage of a drug as a *regulation feedback*, a new therapeutic protocol, a new equipment and a new definition of the medical processes can be considered as an *evolution feedback*.

From the point of view of the emotional influence, referring to the previous chapters, if we want to improve the waiting experience in the waiting areas, we could:

- define some kind of measure (f.i., a questionnaire, to be transformed in a numeric score);
- use the questionnaire to measure the actual situation (initial results);
- define a numeric goal (what score would we reach?);
- improve the waiting area (f.i. by introducing paintings, changing colours and so on) (regulation feedback);
- re-apply the measure with the questionnaire;

- compare the results with the previous ones and with the goals: if the score is better we can confirm our improvements (and spread them over other waiting areas); otherwise, we can return back and try other kind of improvements;
- if the improvements show better results, we can change our goal, and try other actions (f.i. introducing some kind of "book crossing" or puzzles, or music), and repeat the process;
- the experiences teach us new "revolutionary" actions (e.g. no waiting rooms, but a table reserved in an internal bar, with some devices able to inform the moment in which to go), we can apply an "evolution feedback", following the same measurement approach and making it followed by further "regulation" feedbacks.

The evolution of a system is an improvement that will be maintained in the future behaviours and new basis of further progresses. Its goals are, in the order,

- *effectiveness*: to better restore the health of a patient;
- *efficiency*, to do it in less time;
- *cost*, to get the result reducing the expenses.

As clarified in the whole text, we do not cope with medical aspects and refer only to possible design actions. We consider the actions carried on in the health care environment following the emotional design as an *evolution feedback*. Nevertheless, the methodological aspects of our proposals are quite similar to what is usually applied for the medical activity.

7.2 Measures

When we change some elements of the health care process (including the emotional design actions), we have to verify the effectiveness of our actions. That means we need quantitative measures to compare the parameters characterizing the process, before and after the changes.

There are established statistical practices that are applied to evaluate the effectiveness of therapies in the medical field, for example, for drug testing.

The statistical approaches for evaluating that kind of changes are typically defined as *single-blind control procedure*, *double-blind control procedure* and *triple-blind control procedure*. Referring, as a clean example, to the drugs test procedures, the three approaches work as follows:

- *single-blind control procedure*: two samples of homogeneous patients are taken under observation; the drug under test is given to the former group, and a substance pharmacologically inert (*placebo*) to the latter; none of them knows whether they are taking the drug or the *placebo*, in order to avoid psychological biases affecting the results;
- *double-blind control procedure*: as above, but not only the patients, but also the doctors are not aware on who is taking what; this is because the behaviours of the

doctors can be influenced by their expectation, and this can in turn influence the patients;

- *triple-blind control procedure*: as above, but also the statistical analyst processing the data does not know who assumed the drugs and who the *placebo*.

The double-blind approach is mandatory for testing drugs.

The evaluation of the impact of new medical protocols cannot use the double-blind approach, because doctors have the evidence of what they are doing. In this case, the single-blind approach can be applied.

In the case of changes in the environment done according to the principles of the emotional design, the patients have the complete evidence of the changes: more, is just this perceptual evidence to their senses, that rises emotions. So, none of the three approaches can be applied: What we can do is:

- take under observation two homogeneous groups of patients, in two different environments: one in a "traditional" health care centre and the other in an "emotionally improved centre;
- define measurable relevant medical indicators;
- apply the same medical procedures;
- run the measures in both environments and compare them.

So, we could:

- select two different centres;
- select the same pathology and treatment, well defined and standardized, for example, *paediatric appendectomy* with traditional surgery;
- measure the duration of the hospitalization in the two centres and compare the results.

Why *paediatric appendectomy* and not simply *appendectomy*?

Because we need to have as much as possible homogeneous samples, and, due to the fact that this kind of surgery has a long story and fixed protocols, we are just measuring the speed or healing after the operation, and the age could make the difference. In general, such an operation requires 5–10 days; a proved reduction of this duration would allow to confirm the impact of the environment design on the time (and then on the costs).

The same could be done for many specific reasons of hospitalization, collecting data to support and measure the impact of emotional design in the health care environment.

There are many situation in which a similar measure could be difficult: for example, in case of respiratory insufficiencies, heart attacks, hypoglycaemic crises or other similar events, not only the causes can present many differences, but also the severity level can span on a wide range, as well as the general patients' conditions.

Nevertheless, once we verified the positive influence of the emotional design interventions for the cases easier to be measured, we can assume the general validity for any kind of hospitalization.

We stressed the need to compare two environments, one "traditional" (i.e. comparable with the situation before the application of improvements) and the other in which emotional design interventions have been carried on. It is not so mandatory; more, it is not the best way.

If in a centre statistical measure on the past is available, that can be the basis for a comparison, and, if a centre wants to improve some areas, and data are not yet available on the past, the first thing to do is to start with the definition of a set of measures and with the collection of those measures before the application of the changes: those measures will be the basis for the comparison with the future performances.

Once the centre is organized so to be able to do continuous measures, the actual one will be the basis for verifying future changes and then the future improvements.

7.3 Which Measures?

Literature is very rich of indication about measures for the evaluation of the "quality" of the performances of a health care centre.

Many reports provide snapshots of measure to compare the performances in different centres (or regions), but this kind of information (and indicators) are not relevant to our purpose: we need indicators useful for measuring the improvements in the performances related to changes in the environment.

It is quite difficult to define a set of indicators to be measured, valid for any environment, for any health care centre and for any kind of disease. Moreover, a high score in some desirable indicators could be in contrast with the score of other desirable indicators.[1]

Moreover, some indicators that seem desirable could be misunderstood: ten minutes spent in a boring dreary waiting room could be less acceptable than half an hour in a pleasant and distracting area.

For example, in Table 7.1 (Ioan et al., 2012), the indicators should not be considered as measures to be related to specific goals, but just the basis for further elaboration.

For example, most of the indicators related to the *axis efficiency* seem related to economic aspects, but the reduction of the costs could affect the clinical performances. The same consideration applies for the *financial resource management* or for the *focus on the personnel*. Again, we already discussed the relative value of *waiting time* for the *focus on patient* axis.

So, it seems clear that it is not possible to define a universal set of indicators to be measured, to verify the effect of emotional design intervention in the health care

---

[1] There is an Italian proverb saying that "the pitiful doctor makes the purulent sore" (originally, as reported in "Raccolta di proverbi toscani di Giuseppe Giusti" edited by Gino Capponi in 1853, *Il medico pietoso fa la piaga purulenta*). This proverb shows the contraposition between the "customer satisfaction" of reducing pain and the effectiveness of the treatment.

Table 7.1 Examples of indicators usable for an accurate survey on hospitals performance, per processes and strategic axes (Ioan et al., 2012)

Axis	Processes	Indicators
Clinical efficiency	– Care process suitability – Care process compliance – Care process outcomes	– Rate of re-hospitalizations – Rate of mortality – Rate of complications; – Average length of hospital stay
Efficacy	– Using available technology for as much care as possible – Using available technology for the best care services ever possible	– Average length of hospital stay – Average cost – Rate of utilization of the existing technology – Rate of beds occupancy
Financial resources management	– Using available financial resources – Identifying the means of efficient and efficacy resources allocation	– Budget flow compared to approved budget – Emergency services expenses – Hospitalization services expenses – Personnel expenses – Goods and services expenses – Medicine expenses – Average cost per day of hospitalization
Focus on personnel	– Practice environment – Acknowledgement of individual needs – Health promotion activities – Adequate payment – Development (continuous education) – Personnel satisfaction	– Rate of absenteeism – Rate of resignations/transfers – Average payment – Number of specialization courses – Personnel perception
Social accountability and reactivity	– Integration within health system – Integration within community – Orientation towards public health – Access – Continuity – Health promotion – Equity	– Percentage of counselled patients – Percentage of patients with GP/specialist referral – Percentage of patients recommended for discharge – Percentage of patients transferred to other health units
Safety	– Patient safety – Personnel safety – Medical environment safety	– Rate of nosocomial infections – Rate of accidents – Rate of complications

(continued)

Table 7.1 (continued)

Axis	Processes	Indicators
Focus on patient	– Respect for patients – Confidentiality – Communication – Freedom in choosing the physician – Patient satisfaction	– Waiting time – Percentage of patients informed – Patient perception

environments, and every department have to choose the proper measures, according to its specificity.

Nevertheless, we can do the following general consideration:

- some measures are related to "physical" entities (e.g. length of hospital stay, number of some kind of events and so on); among them the measures related to the time have particular relevance (and generality) and are typically easy to be done; time is often related to the effort done by people (patients and doctors and staff) in the healing process, and it is in some way related to the efficiency of the process, both "in the small" (the time to do a medical examination from acceptance to obtaining the final report, measured in hours, or simply the waiting time for a visit, measured, hopefully, in minutes) and "in the large" (the time for the whole care process, since the acceptance to hospital till the discharge);
- some measures are related to the perception of the stakeholder in general (patients, doctors, nurses, staff, relatives, etc.); often such measures tend to verify the satisfaction of a person, status that is heavily dependent on emotional aspects; in this case, we can use techniques such as interviews, questionnaires, and in general procedures related to Kansei engineering, but we can also use some kind of "circumstantial indicators": the walking speed, the "number of smiles" and so on;
- very specific punctual measure can be done, according to the characteristics of the single process: remember, for example, the experience done at the Pausilipon Children Cancer Hospital in Naples, already reported, referring about the measure of the percentage of sedated children during the NMR exams, and the action that makes it fall from about 40 to 2% (a similar case has many impacts: less invasive actions on the children, satisfaction for the parents, less costs, less time).

The indicators we choose for our measurements should not be simply related to the goals: in fact we propose interventions in order to change the performances, and the measure of the results, so, beside measures on the final performances, it is important to measure also the changes in the "input conditions". For example, if we want to measure the patients' satisfaction and use as one of the indicators their smiling, we should measure also the smiling of doctors and nurses, that possibly affect significantly the final result.

Referring to an experience related to emotional design, recently an architect refurbishing a waiting room required us some hints about the lighting bodies. He has

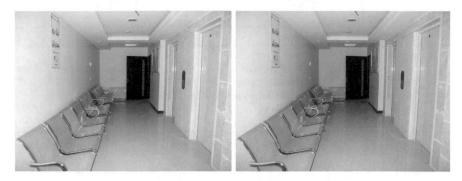

Fig. 7.2 The same waiting room under different lighting conditions (The pictures are just an example and do not refer to a real situation. The original photo, here digitally changed for exemplification purposes [author, Hamidrezaglx], is available on commons.wikimedia.org, under the Creative Commons Attribution-Share Alike 3.0 Unported license). The difference can be perceived just on the digital copy of this book, in colour. *Source* By Authors

shown two photos of a wall, taken when lightened with two different lamps: the former seemed to be lit by a warm setting sun, the latter by a north-facing sky in a cold day (Fig. 7.2).

Our hints were not only to choose the former lighting source, but also to measure the colour temperature and to define how to measure the emotional effect of it.

Many of the effects of the proposed actions would be an improvement of the healing process. In general, the costs of emotional design interventions are marginal in respect to the cost of the health care services; moreover, the "worth" of the health is difficult to be measured economically. The same apply to the emotions: let us suppose that a good lighting will provide a better emotional status, without influencing the healing process[2]: what is the "worth" of an hour of joy?

7.4   How to Measure

A measure is simply an action collecting data. Data are a representation of the real world. In order to better understand the information often hidden within the collected data, a proper representation will help.

For this purpose, we recall shortly the recommendations by K. Ishikawa (1989).

---

[2] But we strongly support that it will have a positive impact on healing.

7.4.1 The Seven Tools of Ishikawa

While a control system applies feedback actions through automatic devices, the improvements of the processes (including the emotional design actions in a system such as a health care environment) are performed by humans: the designer gets some information from the environment, has in mind a (possibly partial) model of the organization, defines changes suitable to improve the performances of it, applies them and finally verifies, again collecting information, the results.

So, we can implement the feedback mechanism by collecting data, defining goals, implementing the feedback actions and verifying them.

Ishikawa indicated seven relevant tools in collecting data, interpreting the measures and so verifying results.

The seven tools are:

- Data collection sheets;
- Histograms;
- Pareto diagrams;
- Pie charts;
- Control charts;
- Correlation diagrams;
- Fish-bone charts.

In the following, we will shortly discuss each of them; the examples we provide are not real, and exaggerated, in order to emphasize their use.

7.4.1.1 Data Collection Sheets

Obviously, if we want to process some information, we must collect them. What we have to decide is:

- Which data?
- How?
- Where?

7.4.1.2 Which Data?

It depends on our goal. As said before, the measure of the effectiveness of the healing process could be simply measured through the recovery time from an illness or the percentage of successful cases or something similar. But in some cases, we can consider a causal chain, being interested in the intermediate steps (we will better examine the case speaking about the fish-bone diagrams). Let us suppose that we consider the following chain (Fig. 7.3).

In other words, after a visit, the doctor will provide the patients with rules to be followed, such as:

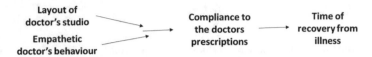

Fig. 7.3 A possible chain of cause-effect, we would try to verify. *Source* By Authors

- get some pills twice a day;
- avoid alcohol;
- avoid some kind of food;
- and so on.

We know that many patients tend to adjust by themselves the indications, forgetting (or voluntarily avoiding) the pills, considering non-relevant to drink very few quantities of spirits and so on.

In this case, we could be interested not only in a data sheet collecting information such as therapy start time—therapy end time but we need to invent some way to measure the compliance to doctor's prescriptions: Percentage of infringement of the rule 1—infringement degree (low–medium–high); Percentage of infringement of rule 2—and so on.

In this case, we have to invent the way to get that information and the scale allowing us to transform the information into numbers.

Following the example, we could record weekly the data as follows (Table 7.2).

Of course, we should use comparable measures in different cases: in the previous example, we observed 12 infringements over 3 rules for 7 days; we could consider a Infringement Index II as

II = (Total Infringement Score)/(Maximum Possible Infringement Score × Number of Days)

In the example,

- the Total Infringement Score is 12;
- the Maximum Possible Infringement Score is the maximum degree (indicated with the value 3) multiplied by the number or rules (3), i.e. 9;
- the Number of Days is 7.

So

Table 7.2 A possible table recording the weekly violations on the doctor's prescriptions: the values 1, 2 and 3 are the numeric expression of the judgement *low, medium, high*

Prescription	M	Tu	W	Th	F	Sa	Su	Score
1. Get pill × twice a day	–	–	–	1	1	–	3	5
2. No alcohol	–	–	–	1	1	1	1	4
3. No fried food	–	–	–	–	–	3	–	3
Total	–	–	–	2	2	4	4	12

Table 7.3 A possible improvement of the evaluation of the violations

Prescription	M	Tu	W	Th	F	Sa	Su	Weight	Score
1. Get pill × twice a day	–	–	–	1	1	–	3	2	10
2. No alcohol	–	–	–	1	1	1	1	1	4
3. No fried food	–	–	–	–	–	3	–	0.7	2.1
Total	–	–	–	2	2	4	4		16.1

$$II = 12/(9 \times 7) = 0.19$$

In such a way we can compare different situations: a Total Infringement Score equal to 24 seems to be twice in respect to the previous example, but, if applied to 7 prescriptions, it will give:

$$II = 24/(7 \times 3 \times 7) = 24/14 = 0.16.$$

Being the measures arbitrary, each health care centre could define a proper evaluation method.

For example, it is possible to include in the computation some weight of the various rules. Referring to the previous example, we could consider as severe a infringement about the pills, less severe the infringement about the alcohol and scarcely severe the infringement about fried food. So, we could consider increasing of the 100% the influence of the first prescription and decrease of the 30% the influence of the third (Table 7.3):

In this case, the Infringement Index should take into account the weights in computing the Maximum Possible Infringement Score:

$$MPIS = 3 \times (2 + 1 + 0.7) = 11.1$$

making II bigger:

$$II = 16.1/(11.1 \times 7) = 0.21.$$

Moreover, higher difficulties could rise when we try to measure the "empathetic doctor's behaviour". We have again to invent some "metrics": for example, we can consider three aspects:

- the general environment;
- the doctor's studio layout;
- the doctor's behaviour.

General environment. We could score it according to the number of taken actions (walls colours, floors colours, furniture, paintings, etc.); for example, we could consider the possible actions examined and evaluate the percentage of them really implemented. In Chapter 3, we listed the possible actions we can do in the general environment (reception, waiting rooms and so on): in a specific case, we could evaluate how many of those actions have been taken.

Roughly, if we listed 10 possible actions and 5 of them have been implemented, we could consider a General Environment Score:

$$GES = 5/10 = 50\%.$$

Of course, we could improve such an evaluation weighting the different parts of the environment (are waiting rooms more relevant of the reception?) or the extent of the actions (are books and paintings in waiting rooms more relevant than the simple change of wall colours?).

Doctor's studio layout. The same apply to the doctor's studio, possibly taking into account also the position of the patient in respect to the doctor. The same considerations of the previous point can apply to build a Doctor Studio Score (DSS).

Doctor's behaviour. The capability of listening, the way of speaking, the way the doctor looks to the patient, the smiles, and many other elements difficult to measure could be relevant. We can approximate all those elements with the visit time, considering that as long time takes the visit as more attention will be felt by the patient. Of course, the evaluation should be transformed in a percentage in respect to a standard evaluation. If we consider a standard visit for a specific purpose as taking 20 min, a visit taking 30 min will be scored as

$$\text{Doctor's Behaviour Score} = DBS = (30 - 20)/20 = 0.5.$$

This measure can range from -1 [no visit: $(0-20)/20$] to any positive value [a visit lasting 2 h will be scored as $(120-20)/20 = 100/20 = 5$].

All the measures we presented here are totally arbitrary and are very difficult to be compared with similar measure in other environments, but what is relevant is the constancy of the measuring approach in a single environment. We must remember that our goal is to orient feedback actions to improve the performances of the whole environment; if we can verify that some actions in changing the emotional content of the waiting rooms improve the healing process, it is not relevant how we measured such actions. It is important only that the measures are stable, consistent and in some way related to the phenomenon they want to capture.

7.4.1.3 How to Collect Data?

Of course, the best way to collect data should be to do it automatically. Among the many possibilities: if each step in the journey of a patient is tracked by a computer application (ruling each bureaucratical action such as check-in, fix a visit, a payment and so on), the related information is already in a computer. In general, those information are not on the same computers as for clinical data, but it is possible to extract them automatically.

It is not possible to provide general indication on how to collect data, large variety. We can observe in any case that there are two extreme levels we can move between:

- the manual one, most expensive and time consuming, but nevertheless effective and useful;
- through ad hoc automated applications; for example, simple applications could be based on RFID (Radio Frequency Identifier) sensor, i.e. very cheap technologies consisting in small circuits able to catch and record information from antennas (again cheap) automatically or manually activated; as we are aware of the small "stamps" attached to many goods against shoplifting. The same technology can record univocally events, parameters and measures.[3] Of course, specific application must be implemented ad hoc. Again, some visual recognition applications could "measure" the "quantity" of "smiles" (as done in some artistic operations).[4]

7.4.1.4 Where to Collect Data?

The answer is trivial: on media able to be read by a computer. If we have some automated mechanism to collect data, those will be necessarily recorded on such a kind of medium.

When, as more common, data are collected and recorded manually, a simple spreadsheet will be mandatory: simple for recording, excellent for building many kinds of representation and excellent for flexibility.

7.4.1.5 Histograms

A histogram is a graphical representation of a phenomenon, in which each measure is represented as a column whose height is proportional to the measure. They put easily in evidence the differences among the values. An example is given in Fig. 7.4, representing the distribution of the cancer typology in Italy in 2020 (Table 7.4).

In Fig. 7.5, it is quite evident (more evident and easier to be observed, in respect to the table) that there are large differences among the data, and this fact suggests splitting the data, and possibly to re-organize them.

---

[3] For example, a simple RFID applied to a patient for a cataract operation could be perceived by a small antenna applied to the bed in the operating room will be able to record automatically the duration of the operation; knowing that an operation with the recent technologies takes approximately 10 min, and that the preparation of the room and of the patient takes at least twice, the measure of the total required time could be useful to verify improvement in efficiency and in costs.

[4] In 2010 three artists, Julius von Bismarck, Benjamin Maus and Richard Wilhelmer, created in Lindau (Germany) a big "smile" on a high lighthouse; the structure allowed the lamps representing eyes and mouth to rotate, the position of the lamps was decided automatically on the "average" mood of the population, according to the recordings done by many webcams distributed in the town, and observing the expressions of the people. The recordings were done respecting the privacy, and just interpreting how many people were smiling. This demonstrates how easy could be the evaluation of the "smiles" in a health care centre. The images are taken from YouTube, *Monumental Interactive Smiley: Fühlometer* by Vernissage TV.

Fig. 7.4 An artistic installation in Lindau (Germany), representing in real time the mood of the population, according to what distributed webcams were observing in the town. *Source* From youtube https://www.youtube.com/watch?v=nBstJ6_HMac—referred

For instances, we can obtain the histogram only for the females (Figs. 7.6 and 7.7).

The examples till now provided allow us to understand the relevance of the histograms in driving our attention towards some kind of phenomena. Let us present some (non-real) histogram referred to the information devoted to the evaluation of the effects of emotional design, towards the implementation of some kind of feedback.

Let us suppose that two hospitals provide cares for the same disease, and that and both record the number of days between check-in and discharge for each patient, according to Table 7.5.

We are interested in observing both, because one of them put in place some actions to improve the performances, since mid-October of that year. We can build the histogram of both, as shown in Fig. 7.8.

It is quite difficult to verify some different behaviour, and in particular there is no evidence that the action undertaken by the first hospital had some effects.

We can add the trend lines on the histograms, as follows (Fig. 7.9).

The trend lines show clearly the reduction in healing time by the first hospital, in respect to a static situation for the second.

So, despite the difficulty in interpreting the data, the trend lines suggest us to do further analysis: we can split the data related to the first hospital into two series, measuring the average value for each period (Fig. 7.10):

The average lines of the two periods as shown in the last figure indicate 7.07 days for the first period and 5.7 days for the second, with an improvement of

$$(7.07 - 5.7)/7.07 = 19\%.$$

Histograms help us in finding the relevant information, without a deep knowledge of statistical analysis.

Table 7.4 Distribution of cancer typology in Italy (2020) (*I numeri del cancro in Italia*, Intermedia editore, 2020, based on data from AIRTUM)

	Male		Female		Total	
	N	(%)	N	(%)	N	(%)
Superior aero digestive tracts	7276	3.7	2580	1.4	9856	2.6
Esophagus	1710	0.9	684	0.4	2394	0.6
Stomach	8458	4.3	6098	3.4	14,556	39
Colorectal	23,420	12	20,282	11.2	43,702	11.6
Liver	8978	4.6	4034	2.2	13,012	3.5
Pancreas	6847	3.5	7416	4.1	14,263	3.8
Gallbladder and biliary tract	2400	1.2	3000	1.7	5400	1.4
Lung	27,554	14.1	13,328	7.3	40,882	10.9
Melanomas	8147	4.2	6716	3.7	14,863	4
Mesothelioma	1523	0.8	463	0.3	1986	0.5
Breast			54,976	30.3	54,976	14.6
Ovary			5179	2.8	5179	1.4
Uterus cervix			2365	1.3	2365	0.6
Uterus body			8335	4.6	8335	2.2
Prostate	36,074	18.5			36,074	9.6
Testicle	2289	1.2			2289	0.6
Kidney, urinary tract	9049	4.6	4472	2.5	13,521	3.6
Bladder	20,477	10.5	5015	2.8	25,492	6.8
Central nervous system	3533	1.8	2589	1.4	6122	1.6
Thyroid	3333	1.7	9850	5.4	13,183	3.5
Hodgkin's lymphomas	1222	0.6	929	0.5	2151	0.6
Non-Hodgkin's lymphomas	7011	3.6	6171	3.4	13,182	3.5
Multiple myeloma	3019	1.6	2740	1.5	5759	1.5
Leukemia, all of them	4738	2.4	3229	1.8	7967	2.1

7.4.1.6 Pareto Diagrams

Many complex phenomena present a behaviour well described by a mathematical law known as *power law* or *Pareto distribution* (from the name of the scientist Vilfredo Pareto, who studied many social and economic phenomena following this law). Without entering in mathematical details, the law expresses the concept that in many complex systems in which various elements can cause a specific effect, roughly, the 20% of the causes are responsible of the 80% of the effects.

The discovery that a phenomenon follows a Pareto distribution allows us to concentrate the actions towards the few causes more influencing the results.

For this reason, it is relevant to draw a Pareto diagram. We can do it by operating as follows: we order the data by decreasing value, then we calculate the incremental

Table 7.5 Days to recover in two hospitals

Initial date	Therapy days	
	Hospital 1	Hospital 2
6 September 2017	8	7
13 September 2017	7	7
21 September 2017	7	7
27 September 2017	6	6
28 September 2017	8	7
29 September 2017	7	7
30 September 2017	6	6
30 September 2017	8	7
30 September 2017	5	5
03 October 2017	9	6
07 October 2017	7	7
12 October 2017	8	6
13 October 2017	7	7
14 October 2017	6	8
16 October 2017	6	6
16 October 2017	7	5
16 October 2017	5	8
16 October 2017	6	7
21 October 2017	5	6
25 October 2017	6	7
30 October 2017	7	8
5 November 2017	4	6
6 November 2017	6	6
7 November 2017	5	7

percentage, as the number of cases increases (which percentage is constituted by the first, which by the first two, which by first three and so on).

Looking at Fig. 7.7 of this chapter, we can observe that some cancer type is largely diffused, while some other seems to be less frequent. We can redraw the figure adding a curve representing the incremental percentage of the cases (Fig. 7.11).

From the figure, we can see that the breast cancer corresponds to the 32% of the cancer cases (scale on the right), that the breast and the colorectal cancers together represent the 44% of the cases, and that the first 6 cases (constituting the $6/22 = 27\%$ of the "causes") are responsible of the 67% of the illnesses.

Of course, we can use the Pareto diagrams to understand which intervention we have done in a department have been more influent on the emotional status of the patients.

Suppose that we have introduced some new elements in a waiting room:

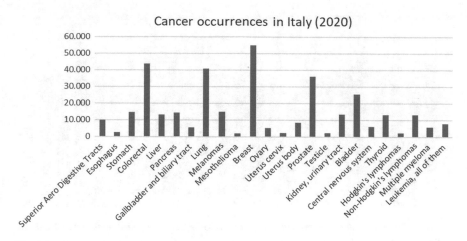

Fig. 7.5 The histogram of the data reported in Table 7.4. *Source* By Authors

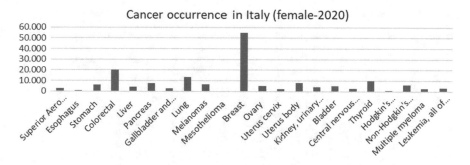

Fig. 7.6 Distribution of cancer typology in Italy (2020) for females. *Source* By Authors

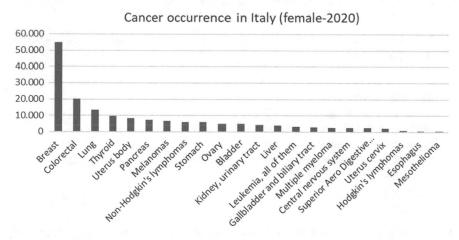

Fig. 7.7 Distribution of cancer typology in Italy (2020) for females, reordered by occurrences. *Source* By Authors

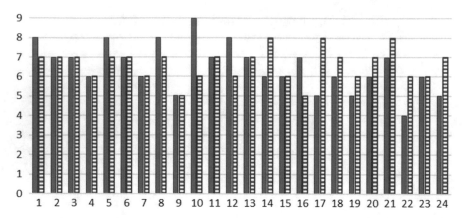

Fig. 7.8 The histograms of the number of days required for a complete healing for a disease in two hospitals. *Source* By Authors

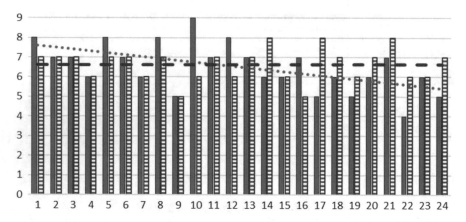

Fig. 7.9 The same histograms with the trend lines: dotted for the first hospital and dashed for the second. *Source* By Authors

- paintings on the walls;
- newspapers and magazines;
- information booklets on diseases and therapies;
- books with short novels;
- booklets to write impressions and thoughts available for other patients;
- short videoclips;
- crossword puzzles;
- background music;

and suppose to collect the points of view of many patients about the appreciation of the various initiatives with respect to the ability to distract and therefore not to feel the annoyance of waiting.

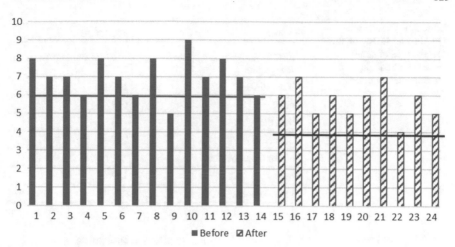

Fig. 7.10 The histograms of the first hospital in the two periods: before and after the improvements. *Source* By Authors

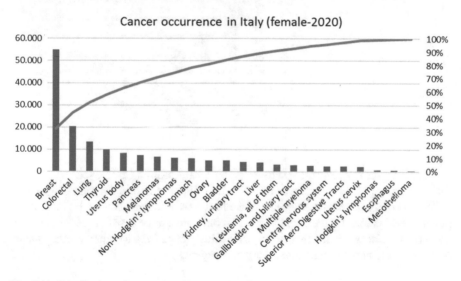

Fig. 7.11 Distribution of cancer typology in Italy (2020) for females, reordered by occurrences: the line represent the incremental cumulative percentage of the cases. *Source* By Authors

We could, for example, organize some questionnaires, asking to which extent each element contributed to accept the waiting, in a form as the called *Likert scale*:

☐   ☐   ☐   ☐   ☐

Irrelevant   Scarce   Influential   Relevant   Decisive

Table 7.6 Results of a
hypothetical survey on the
effectiveness of some
elements in reducing the
stress in waiting areas

Element	Total score
Paintings	2695
Newspapers	1877
Crosswords	1080
Music	643
Read thoughts	403
Videoclips	217
Information booklets	182
Novels	132
Write thoughts	92

Fig. 7.12 The Pareto
diagrams of the survey
described in Table 7.6.
Source By Authors

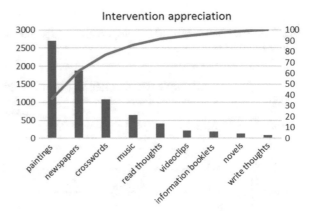

Let's imagine to obtain the results as in Table 7.6, summing all the scores collected for each element from which we can draw the following Pareto diagram (Fig. 7.12).

We can observe than the presence of the first three elements covers the 77% of the positive evaluations by the patients. So, not only we can be satisfied the undertaken actions, but also we learn that, when extending the actions to other waiting areas, those three elements will be mandatory and urgent.

7.4.1.7   Pie Charts

A pie chart divides a circle into sectors whose angular extension is proportional to the percentual weight of the measure.

For example, in Fig. 7.13, it is represented the pie chart referred to the results of the survey supposed in Table 7.6.

The chart does not show the evidence of the values, but emphasizes the weight of each element: from the figure, we understand the relevance of the paintings on the other aspects, and on its basis we could discard other actions such as videoclips

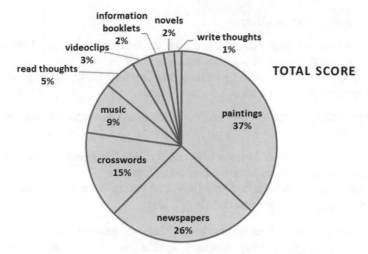

Fig. 7.13 The pie chart of the survey described in Table 7.6. *Source* By Authors

and following. Moreover, we could take into account that the expenses for paintings are "one shot" costs while the others require maintenance or handling current costs; the chart put in evidence the role of the paintings in reducing stress, and it could be decided to start just with this only element.

7.4.1.8 Control Charts

A control chart describes the behaviour of some phenomenon in the time.

For example, Fig. 7.14 provides a hypothetical behaviour of the average waiting time at the initial registration, expressed in minutes: each measure refers to a specific day, within a range of 3 months.

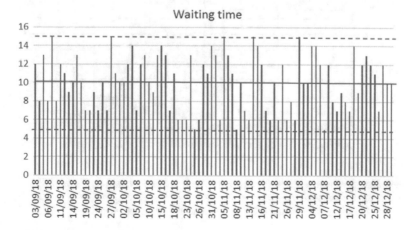

Fig. 7.14 Average waiting time for the initial registration. *Source* By Authors

From the figure, we see that:

- the behaviour is stable, i.e. there are high values and low values distributed in such a way that we cannot find trends of any kind;
- the average of the average daily waiting time is a little bit higher than 10 min;
- the average values vary between 5 and 15 min.

Different phenomena have different behaviours, in general well represented by a control chart:

- the example provided in Fig. 7.14 is related to a stable phenomenon; the values are randomly distributed around a horizontal line;
- epidemic diseases show a fast growth of cases in the time, till a maximum, and then a, generally slower reduction (Fig. 7.15);

- some phenomena, such as the number of elements integrated on semiconductor microchips, are growing exponentially; in those cases, the control chart should use a logarithmic scale (see Fig. 7.16).

Returning to the average waiting time of Fig. 7.14, our goal is to improve the process, and, for a stable behaviour, we should consider two elements: the average value and the width of the variability band between minimum and maximum; we could have to goals:

- increase the "quality", and in this case it means to reduce the average waiting time;
- reduce the variability of the waiting time, that is to reduce the width of the variation band.

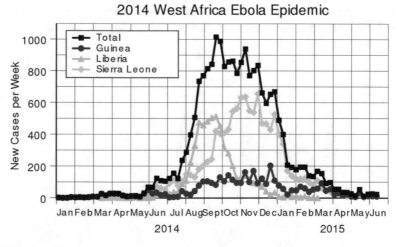

Fig. 7.15 New Ebolavirus cases per week in some African countries. *Source* WHO; From: https://commons.wikimedia.org/wiki/File:2014_West_Africa_Ebola_Epidemic_-_New_Cases_per_Week.svg?uselang=it—Available for reuse if unchanged, as we have done

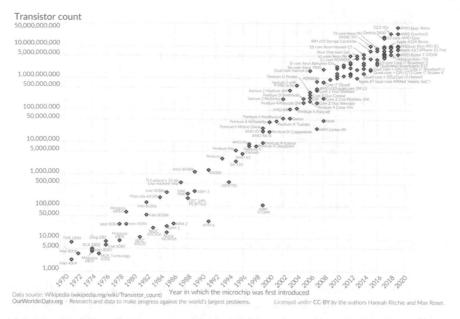

Fig. 7.16 A logarithmic graph showing the timeline of how transistor counts in microchips are almost doubling every two years from 1970 to 2020; Moore's Law. *Source* Max Roser, Hannah Ritchie—ourworldindata.org; From https://commons.wikimedia.org/wiki/File:Moore%27s_Law_Transistor_Count_1970-2020.png17. Licensed under the Creative Commons Attribution 4.0 International18. Available in Wikimedia Commons, free use if the source is acknowledged, as we have done

When we want to improve the performances in a similar case, we need to investigate on the possible cause-effect relationships. We will discuss the topic in the paragraph on fish-bone charts.

7.4.1.9   Correlation Diagrams

If some phenomena are a cause of others, it can be revealed by a correlation diagram, plotting the measures of the two phenomena for each of the observed instances.

A very simple example is given by a diagram plotting the weight of an object in respect to its volume (Fig. 7.17).

Of course, the trivial example is just the representation of a simple formula referred to the concept of *density*:

$$\text{Weight} = \text{Density} \times \text{Volume}.$$

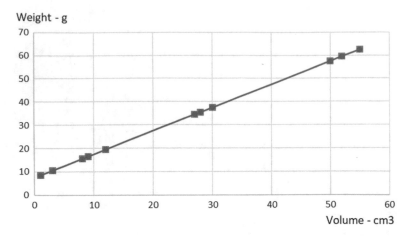

Fig. 7.17 The diagram correlating volume and weight of objects made of the same substance

More interesting is a case in which we do not refer to a law derived by a physical definition, but to not well-known phenomena, for which the various points represented on the diagram have to be interpreted statistically; for example, the diagram of Fig. 7.15 shows that as the income inequality grows, obesity grows.

In this case, we are not facing a "physical law", but just verifying that two phenomena could be related, one as cause of the other.

The correlation can show many different behaviours: in the previous example there was a *direct relation* between the measures (the dependent measure grows proportionally to the independent), but we can observe other cases: in Fig. 7.16, we observe that as the time spent in eating grows, the obesity percentage decreases (*inverse correlation*) (Fig. 7.18).

While the previous examples was presenting a *linear* trend (the ratio between the two measures on the two axis x and y is a constant), the trend in Fig. 7.19 shows that

Fig. 7.18 Obese population percentage (The Spirit Level, Wilkinson & Pickett, Penguin 2009)

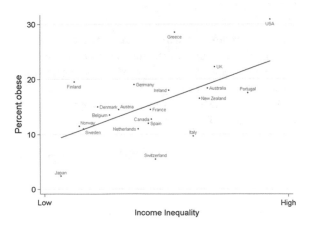

Fig. 7.19 Time spent eating per day versus National Obesity Rate. *Source Organization for Economic Cooperation and Development*; Public data—The same figure can be found in many websites—We referred the original source—We consider "fair use"

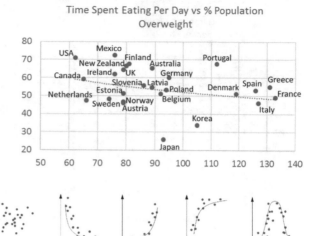

Fig. 7.20 Various typical possible distributions. *Source By Authors*

as the time on axis x grows, the value on axis y decreases, but more and more slowly (possibly a decreasing exponential trend).

Figure 7.17 shows different typical distributions, in which we can recognize different trends (Fig. 7.20).

In each diagram, the dots represent the done measures, while the line represents the supposed behaviour:

a. ascending linear;
b. descending linear;
c. no correlation: casual distribution;
d. decreasing exponential;
e. growing exponential;
f, exponential saturation;
g. Gaussian distribution.

When we plot the measures using a common spreadsheet, it is possible to trace the curve that better approximate the points: for example, Fig. 7.21 shows the relationships between the number of yearly smoked cigarettes and the number of deaths for lung cancer; the dotted line is the curve that best fits the points, and the spreadsheet (in this case Microsoft Excel) provides also the mathematical expression of the line. Without entering in technical details of statistics, the coefficient indicated as R^2 is a measure of how well the line fits the points and can be considered as a guarantee of the correlation: it varies between 0 and 1, being 0 the case of totally unrelated values and 1 totally related values.

The spreadsheets provide in general different choices for a best-fit curve, and the type of it should be carefully chosen according to the distribution of the data and to

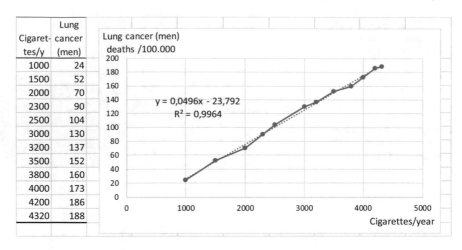

Cigaret-tes/y	Lung cancer (men)
1000	24
1500	52
2000	70
2300	90
2500	104
3000	130
3200	137
3500	152
3800	160
4000	173
4200	186
4320	188

Fig. 7.21 Relationship between smoked cigarettes and lung cancer deaths. (Data adapted from Wikipedia [Italian version, search "Neoplasia"]); By Authors

the kind of the phenomenon. In the abstract example of Fig. 7.22, we have drawn three different best-fit lines:

- the first (equation on the left) is linear and seems to approximate well the points, having a value of R^2 higher than 0.8;
- the second (equation on the right) is an exponential and approximates the point better, being R^2 at 0.91;

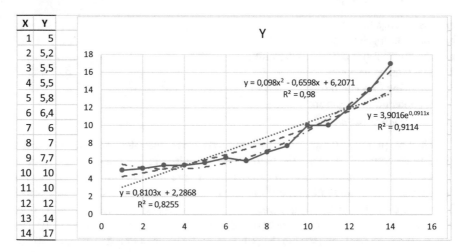

X	Y
1	5
2	5,2
3	5,5
4	5,5
5	5,8
6	6,4
7	6
8	7
9	7,7
10	10
11	10
12	12
13	14
14	17

Fig. 7.22 An abstract example, showing different possible curves fitting with the done measures. *Source* We consider the pictures as for "fair use": they have been found on internet, in the sites of the producers or of the resellers.

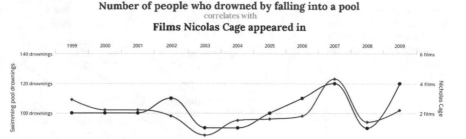

Fig. 7.23 A spurious apparent correlation. *Source* www.tylervigen.com/spurious-correlations

– the third, approximated through a polynomial curve (degree 2, a parabolic behaviour), seems to be the best, reaching the value $R^2 = 0.98$.

If we find some kind of correlation between two physical quantities, we must be careful in interpreting it as a relationship of cause-effect. Sometimes the correlation is the result of a third phenomenon associated to both; in other cases, it could happen simply by chance.

For example, statistics could be built, showing a relationship between the use of a bowler and the life expectancy in the twenties: the relationship is not related to healthy properties of this kind of hut, but to the fact that just the upper class (i.e. richer and able to access to health care) was wearing it.

Again, a very famous case (because quoted in a letter by Gustave Flaubert to his sister) is the relationship between the height of the mast of a sailing ship and the age of its captain: the height of the mast is an indirect measure of the tonnage of the ship, i.e. of the value of the cargo, which is related to the required experience, and then to the age of the captain (Baruk, 1992).

Again, among the odd relationships appearing by chance, we report the following: a comparison in the years between the number of persons who drowned by falling in a pool and the number of films in which Nicholas Cage was starring. The correlations between the two kinds of events seem evident, but of course there is no real correlation at all (Fig. 7.23).

7.4.1.10 Fish-Bone Charts.

A *fish-bone chart* is a tool to investigate (and then model) a phenomenon in terms of cause-effects.

To present how it works, we consider the chart in Fig. 7.24, modelling the leverages we can use for improving the success of a restaurant (the example is very simplified for the purpose of explaining the use the fish-bone charts).

The elements influencing the success of a restaurant are mainly the following:

Fig. 7.24 A simple fish-bone chart modelling the leverages we could use for the success of a restaurant. *Source* By Authors

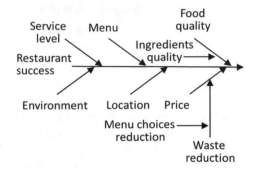

- the quality of the *food* and the *price* for a meal;
- the personality and the variety of the *menu*;
- the level of the *service* by the waiters;
- the characteristics of the *environment* (furniture, layout, walls, etc.);
- the *location* in a specific urban area.

So, we draw an arrow named *restaurant success* and draw other arrows, incident with the first one, each of them representing an element influencing the success.

In this way, we have a list of the elements to be improved in order to reach our goal.

More, we can apply the same process to each of the secondary arrows, indicating, for example, that:

- the *food quality* largely depends on the *quality of the used ingredients* and that implies the fact that we cannot reduce the prices through a reduction of the ingredient's costs;
- the reduction of the *prices* should then pass through a reduction of *wasting*; because a large number of dishes implies a large number of different ingredients, and then the possibility to expire for part of them, the reduction of wasting passes through the *reduction of the number of choices on the menu*.

So, a fish-bone chart helps us in investigating the actions useful for improving the performances of an organization.

Moreover, when we are able to measure the effectiveness of an action, we can also define priorities in the actions to be done.

For example, the pie chart of Fig. 7.13 reflects the effectiveness of some possible actions in reducing the stress in a waiting area; the same can be represented as a weighted fish-bone chart as follows (Fig. 7.25).

The evidence of the different weights drives us in taking the more effective actions.

In the previous chapters, we presented many possible actions we can do through emotional design practices in the health care environments, and we can represent those action in a structured way, organizing them in a fish-bone chart.

We present a general structure of the possible area of intervention in Fig. 7.26,

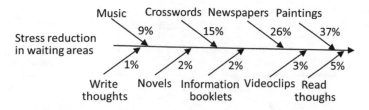

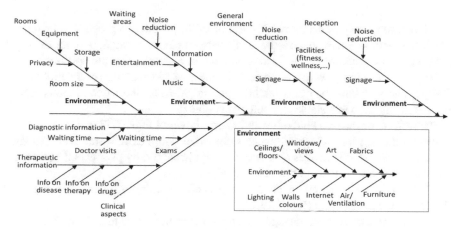

Fig. 7.25 The pie chart of Fig. 7.13 transformed in a fish-bone chart. *Source* By Authors

Fig. 7.26 The fish-bone chart of the possible emotional design actions in a health care centre. *Source* By Authors

but it is just an example: each organization should build its own chart, according to the specific needs and characteristics.

In that figure, the sub-chart **Environment**, common to many possible leverage, has been drawn separately and can apply to all the references in the big chart.

In the following page, a real fish-bone chart investigating the long wait areas in hospital is presented (Shafieyoun, 2016).

7.5 Continuous Improvements

In the 1960s the application of Total Quality to manufacturing processes grew in interest, becoming an important element of corporate reorganization. The application of Total Quality methodologies, born in Japan, spread overall in the world.

The principles behind this approach are not far from the feedback scheme presented at the beginning of this chapter, and, despite the disappearance of the term Total Quality, the approach remains still valid.

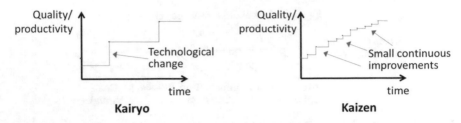

Fig. 7.27 The two approaches Kairyo and Kaizen to the process improvements. *Source* By Authors

The approach required a formal description of the production processes and quantitative measures of some parameters in the various phases along it, in order to make the production structure evolve, also in absence of technological breakthrough.

Two Japanese terms were used to describe the possible evolution of the processes: *Kairyo*, due to technological breakthrough, and *Kaizen*, due to continuous small improvements also in a stable technological context.

The effects of the two kind of evolution are shown in Figs. 7.27 and 7.28.

While the big technological changes are often complex and traumatic for the companies, the small changes cost less, are well absorbed by people and structure, and can reach the same quantitative advantages.

The need to verify quantitatively the results and to continuously improve imposes the so-called PDCA cycle (Fig. 7.29).

So, we can operate small changes in a health care environment, always following this approach:

- *Plan*

    - Focus on a specific area;
    - Identify the involved processes and their structure;
    - Use a fish-bone chart to investigate about the possible required effects and the possible causes, and related intervention;
    - Define which measure can be used to verify the effects, and possibly, to verify the extent of the intervention; identify which among the Ishikawa conceptual tools will be useful.

- *Do*

    - Select a small typical environment to test the solution;
    - Set up the measurement system, and verify to have enough measures to check the changes after the interventions or to have measures for a comparable environment (to provide a comparison basis);
    - Do the proposed changes and start the measurement campaign.

- *Check*

    - Compare the measures obtained after the changes with the comparison basis;

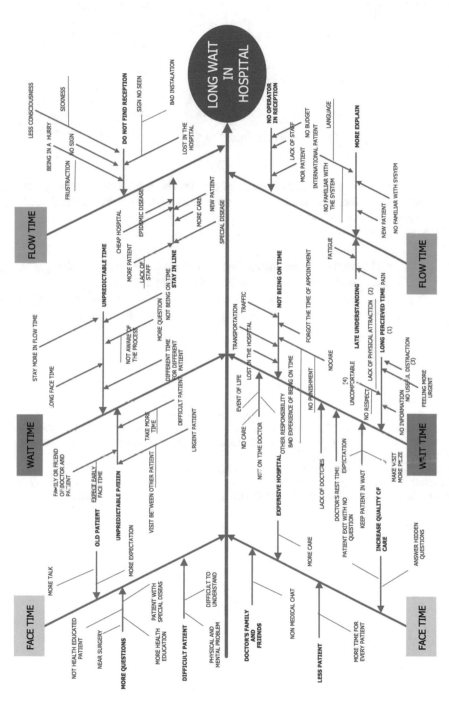

Fig. 7.28 Fish-bone chart analysing the long wait in hospitals. *Source* By Authors

Fig. 7.29 The cycle
Plan–Do–Check–Act.
Source By Authors

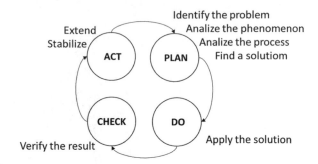

- Evaluate the impact and provide a judgement about the effectiveness of the done action; the main question is: are the costs of the improvements suitable in respect to the evidence of advantages?

- *Act*

  - If the results got though the changes are poor, we have to revise the plan phase, investigating with more accuracy or through alternative paths;
  - If the results are positive, we can extend the actions to all the possible intervention areas, making the changes stable and "institutionalized".

References

Baruk, S. (1992) *Dictionnaire de mathématiques élémentaires.* Seuil.
Ioan, B., Nestian, A. S., & Tiţă, S-M. (2012). *Relevance of Key Performance Indicators (KPIs) in a hospital performance management model. Journal of Eastern Europe Research in Business & Economics.* IBIMA Publishing.
Ishikawa, K. (1989). *Introduction to quality control. Tokyo, 3A Corporation.*
Shafieyoun, Z. (2016). *Affecting emotions through design. How design cam affect the emotions of waiting time in the healthcare centres* (PhD Thesis in Design). Politecnico di Milano.

Chapter 8
A Future Scenario

The health status of the population in more developed countries has substantially improved in recent decades, due to the advances resulting from medical and pharmaceutical research, and the strengthening of national health systems, supported by public expenditure.

According to many studies, the costs of public health are constantly increasing, both in absolute terms and as a percentage of national GDPs, as shown in Fig. 8.1.

Any effort aiming at reducing these costs, without compromising the effectiveness of cares, is crucial for the economy of all the countries. Design research can play a substantial role to tackle this challenge. Given that design is fundamentally a transdisciplinary problem-solving activity, it can bridge the gap among disciplines and suggest new approaches.

Many studies—and a general belief among doctors—underline the role of many factors, normally neglected by standard procedures, that may affect successful treatment, thus supporting the system efficiency and reducing costs. One among these factors is the attitude of patients: their hope for recovery, their confidence in the doctors and their compliance to the required behaviours.

Assuming that the quality of an environment can positively affect the emotional status of a person, a patient; in this case, targeted design actions can be undertaken to reach this goal, thus increasing patients' trust in doctors and in organizations, encouraging their compliance to prescriptions or even inducing a better reaction to therapies.

According to neuroscience and medicine, the psychological status of a person is strongly related to the activation of the sympathetic and parasympathetic nervous systems, both influencing many vital functions and including the immune system.

The same disciplines have highlighted the strong relationship between the central nervous system production of different neurotransmitters regulating emotional processes and the response of the peripheral nervous systems.

According to medical practice, the psychological status of a patient can largely affect both the efficiency and the efficacy of medical therapies.

© The Author(s), under exclusive license to Springer Nature Switzerland AG 2022   139
M. M. Maiocchi and Z. Shafieyoun, Emotional Design and the Healthcare Environment,
Research for Development, https://doi.org/10.1007/978-3-030-99846-2_8

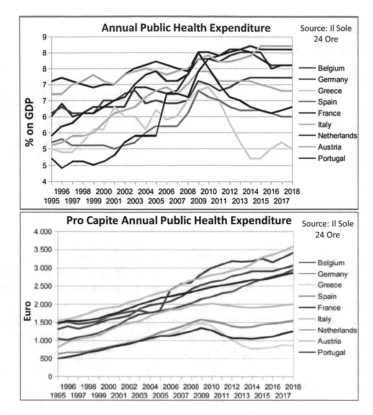

Fig. 8.1 Trends of the public health expenditure for some European countries. *Source* Indicated—
Fair use

Last, according to the discipline of design, the shape, layout and features of an
environment can influence our own behaviour as well as affect our own emotions
and moods.

At the Department of Design of the Politecnico di Milano, studies and field
research have been going on for many years on health care environments, on design
strategies to support medical therapies, on emotional design:

– the Laboratory of Biomedical Sensors and Systems developed innovative methodologies and models for the measurement of biomedical signals and their
  processing in medicine, in assistive technology and ergonomics (Andreoni et al.,
  2014);
– the Laboratory of Innovation and Research about Interiors investigated the role of
  physical environment in non-pharmacological treatment of dementia (Biamonti,
  2018);
– the Colour Laboratory developed theoretical research analysing the relationship
  human/colour and field research on the effects of colour and light in health care
  centres (Calabi et al., 2019);

– the Laboratory of Studies on interaction and perception, undertook studies on the role of emotions in design processes, thus probing the contribution of neurosciences to the discipline of Design (Maiocchi, 2015).

Moreover, a rich and spread knowledge has been acquired in the field by many excellent research groups around the world.

The competence acquired by these groups and the result of their studies can converge into a strong line of research aimed at testing the contribution of health care environment in supporting medical therapies, through an investigation on how design actions in the environment can exert an emotional impact on patients.

It is strong belief of the authors that a worldwide joined research and experimentation on the positive effects in applying emotional design principles for health care environments could lead to relevant results:

– verifying and so defining the best practices for the health care environments;
– stating the best architectural standards to be followed for the new buildings;
– improving the effectiveness of the therapies without extra costs (colouring walls with yellow instead of grey has no extra costs, as for many others design choices), i.e. no extra costs both for capital and operational expenses.

References

Andreoni, G., Barbieri M., & Colombo, B. (2014). *Developing biomedical devices. design, innovation and protection*. Springer.

Biamonti, A. (2018). *Design & Alzheimer. Dalle esperienze degli Habitat Terapeutici al modello GRACE*. FrancoAngeli

Calabi, D., Bisson, M., & Venica, C. (2019). *Design and medical training experimental hypotheses for training in immersive Environments*. 3rd International Conference on Environmental Design.

Maiocchi, M. (2015). *The neuroscientific basis of successful design*. Springer.

Index

© The Editor(s) (if applicable) and The Author(s), under exclusive license 143
to Springer Nature Switzerland AG 2022
M. M. Maiocchi and Z. Shafieyoun, Emotional Design and the Healthcare Environment,
Research for Development, https://doi.org/10.1007/978-3-030-99846-2

Printed in the United States
by Baker & Taylor Publisher Services